World Flags

 Afghanistan

 Albania

 Algeria

 Andorra

 Angola

 Antigua and Barbuda

 Argentina

 Armenia

 Australia

 Austria

 Azerbaijan

 Bahamas

 Bahrain

 Bangladesh

 Barbados

 Belarus

 Belgium

 Belize

 Benin

 Bhutan

 Bolivia

 Bosnia-Herzegovina

 Botswana

 Brazil

 Brunei

 Bulgaria

 Burkina

 Burundi

 Cambodia

 Cameroon

 Canada

 Cape Verde

 Central African Republic

 Chad

 Chile

 China

 Colombia

 Comoros

 Congo

 Congo, Dem. Rep.

 Costa Rica

 Côte d'Ivoire

 Croatia

 Cuba

 Cyprus

 Czech Republic

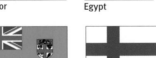 Denmark

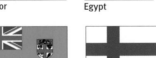 Djibouti

 Dominica

 Dominican Republic

 East Timor

 Ecuador

 Egypt

 El Salvador

 Equatorial Guinea

 Eritrea

 Estonia

 Ethiopia

 Fiji

 Finland

 France

 French Guiana

Gabon

 Gambia

 Georgia

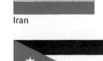 Germany

 Ghana

Greece

Greenland

 Grenada

 Guatemala

 Guinea

 Guinea-Bissau

 Guyana

Haiti

 Honduras

 Hungary

Iceland

India

Indonesia

Iran

Iraq

Ireland

Israel

Italy

Jamaica

Japan

Jordan

Kazakhstan

Kenya

 Kiribati

Kuwait

Kyrgyzstan

Laos

Latvia

Lebanon

Lesotho

Liberia

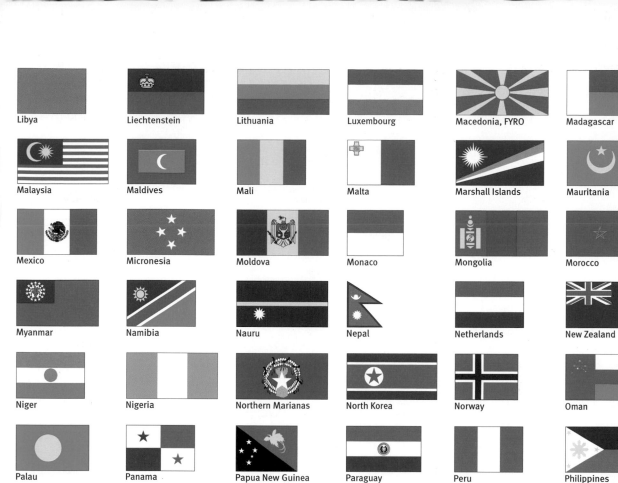

Libya	Liechtenstein	Lithuania	Luxembourg	Macedonia, FYRO	Madagascar	Malawi
Malaysia	Maldives	Mali	Malta	Marshall Islands	Mauritania	Mauritius
Mexico	Micronesia	Moldova	Monaco	Mongolia	Morocco	Mozambique

Myanmar	Namibia	Nauru	Nepal	Netherlands	New Zealand	Nicaragua

Niger	Nigeria	Northern Marianas	North Korea	Norway	Oman	Pakistan

  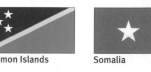 

Palau	Panama	Papua New Guinea	Paraguay	Peru	Philippines	Poland

Portugal	Qatar	Romania	Russian Federation	Rwanda	St. Kitts and Nevis	St. Lucia

St. Vincent & the Grenadines	Samoa	San Marino	Sao Tomé and Principe	Saudi Arabia	Senegal	Serbia and Montenegro

Seychelles	Sierra Leone	Singapore	Slovakia	Slovenia	Solomon Islands	Somalia

South Africa	South Korea	Spain	Sri Lanka	Sudan	Suriname	Swaziland

Sweden	Switzerland	Syria	Taiwan	Tajikistan	Tanzania	Thailand

  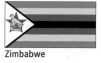 

Togo	Tonga	Trinidad and Tobago	Tunisia	Turkey	Turkmenistan	Tuvalu

Uganda	Ukraine	United Arab Emirates	United Kingdom	United States of America	Uruguay	Uzbekistan

Vanuatu	Venezuela	Vietnam	Yemen	Zambia	Zimbabwe	

OXFORD
Primary
ATLAS

Editorial Adviser
Dr Patrick Wiegand

OXFORD
UNIVERSITY PRESS

Great Clarendon Street, Oxford OX2 6DP

Oxford University Press is a department of the University of Oxford.
It furthers the University's objective of excellence in research, scholarship,
and education by publishing worldwide in

Oxford New York

Auckland Bangkok Buenos Aires Cape Town Chennai
Dar es Salaam Delhi Hong Kong Istanbul Karachi Kolkata
Kuala Lumpur Madrid Melbourne Mexico City Mumbai Nairobi
São Paulo Shanghai Taipei Tokyo Toronto

ISBN 0 19 832076 0 (hardback)
ISBN 0 19 832077 9 (paperback)

Printed in Italy

Acknowledgements

Illustrations by:
Julian Baker p 20; Adrian Barclay pp 31, 35, 41, 51, 54, 55; Mark Duffin pp 7 (compass), 23tl, tr & br, 24 (bricks), 26, 59; Nick Hawken pp 24 (settlements), 32, 33, 36, 37, 38, 39, 42, 43, 45, 46, 47, 48, 49, 52, 53, 55; Tracey Learoyd and Adrian Smith p 20 *and thereafter* (landscape pictograms); ODI p 24 *and thereafter* (population figures); Harry Venning p 60

The publishers would also like to thank the following for permission to reproduce the following photographs:
Alamy pp 20t (Robert Harding Picture Library), 20ct (The Photolibrary Wales), 20c (Geogphotos), 20cb (Worldwide Picture Library), 22tr (Leslie Garland Picture Library), 24t (David Crausby), 24c (Elmtree Images), 26t (David Martyn Hughes), 26b (Gina Calvi), 29ct (Jon Arnold Images), 29cb (Ian Thraves), 39t (Robert Harding Picture Library), 48t (ImageState), 60cr (Steve Bloom Images), 60l (TH Foto), 60r (Guy Somerset), 64tc (Robert Harding Picture Library), 64bl (ashfordplatt); Corbis pp 20b (Chinch Gryniewicz), 22tl (David Paterson), 23tl (Neil Beer), 26c (Jason Hawkes),28t (Martin Jones), 33b (ML Sinibaldi), 36t (Lindsay Hebberd), 36b (Richard Bickel), 42br (Charles Lenars), 52t (Jeremy Horner), 64tr (Ron Watts), 65tr (Richard A. Cooke), 65bc (Galen Rowell), 65bl (Wolfgang Kaehler); Frank Lane Picture Agency pp 29b (Chris Demetriou), 42br (Derek Hall), 43ct (Peter Davey), 43br (David Hosking), 64tl (Minden Pictures); Getty Images/Photographer's Choice p 46l

(James Randklev); Getty Images/Stone pp 24b (Patrick Ingrand), 29tr (Tony Page), 33t, 42t (Will & Deni McIntyre), 43l (Daryl Balfour), 46r, 52b (Pascal Rondeau), 58br; Getty Taxi pp 28b (Richard Cooke), 39b, 47r (B & M Productions), 48b (Tom Bean); Getty Images/The Image Bank pp 47l, 58bl (Image Makers), 64br, 65tl, 65br (Frans Lemmens); Heritage Image Partnership © The British Museum p 28ct (Institution Reference: M&ME, 1939,10-10,93); Powerstock p 28cb (Superstock); Science Photo Library pp 8 (NRSC Ltd), 59l (Earth Satellite Corporation), 59r (Planetary Visions Ltd), 60cl (David Vaughan); © UK Perspectives p 27.

The page design is by Adrian Smith.

The publishers are grateful to the following colleagues in geography education for their helpful comments and advice during the development stages of this atlas:

Jeremy Bullock, Susan Butler, Claire Condie, John Dewis, Tracey Ellis, John Halocha, Joan Huckle, Richard Jefferies, Pat Kelway, Amanda Lightfoot, Trevor Mason, Vanessa Richards, Vicky Stevenson, Emma Wells, Niki Whitburn, Brenda Whittle.

The publishers would also like to thank Phoenix Mapping and Suzanne Williams for their help during the production of this atlas.

2 Contents

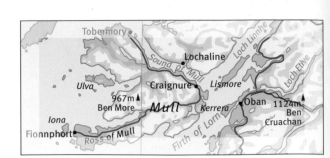

The United Kingdom

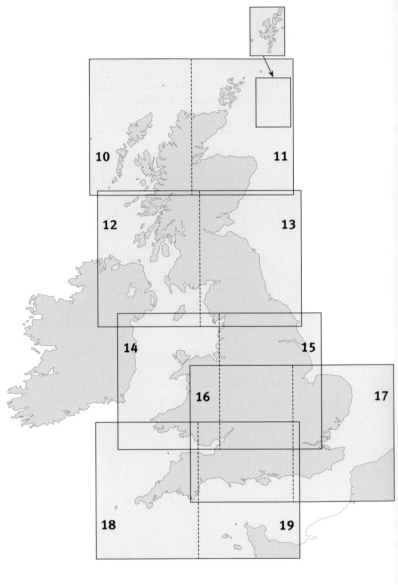

Europe

Asia

Africa

North America

South America

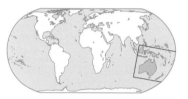

Oceania

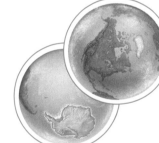

The Poles

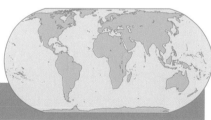

World

4 Atlas literacy

Map language

There are special names for the parts of maps

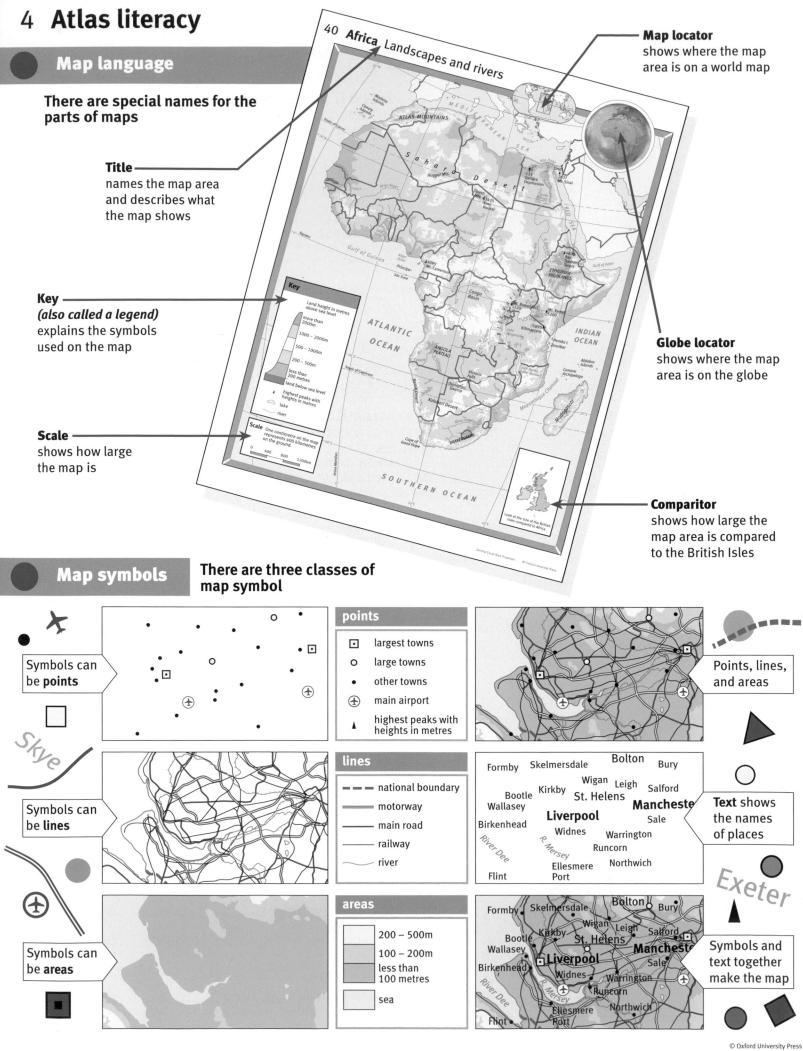

Title
names the map area and describes what the map shows

Key
(also called a legend)
explains the symbols used on the map

Scale
shows how large the map is

Map locator
shows where the map area is on a world map

Globe locator
shows where the map area is on the globe

Comparitor
shows how large the map area is compared to the British Isles

Map symbols

There are three classes of map symbol

Symbols can be **points**

Symbols can be **lines**

Symbols can be **areas**

points
▣	largest towns
○	large towns
•	other towns
✈	main airport
▲	highest peaks with heights in metres

lines
━ ━	national boundary
═══	motorway
───	main road
───	railway
～～	river

areas
	200 – 500m
	100 – 200m
	less than 100 metres
	sea

Points, lines, and areas

Text shows the names of places

Symbols and text together make the map

© Oxford University Press

Type on maps

The way text is printed on maps gives an important clue to what the words mean

Great Britain *Ireland*	islands
UNITED KINGDOM REPUBLIC OF IRELAND	countries
ENGLAND SCOTLAND WALES NORTHERN IRELAND	parts of the United Kingdom
PENNINES GRAMPIAN MOUNTAINS	physical features
Ben Nevis Snowdon	mountain peaks
NORTH SEA *English Channel*	sea areas
Manchester York Dover	settlements

Map abbreviations

An abbreviation is a shortened version of a word or a group of words

Some country names are abbreviated using the first letters of each word

UK   United Kingdom
USA United States of America
UAE United Arab Emirates

R.   River
Mt.   Mount
Is.   Island
Pen. Peninsula

Country names and adjectives

There are patterns in the way some country names make adjectives

Australia	**Australian**	Ireland	*Irish*
India	**Indian**	Poland	*Polish*
Nigeria	**Nigerian**	Sweden	*Swedish*
Zambia	**Zambian**	Turkey	*Turkish*

China	**Chinese**	Brazil	*Brazilian*
Japan	**Japanese**	Canada	*Canadian*
Malta	**Maltese**	Egypt	*Egyptian*
Taiwan	**Taiwanese**	Italy	*Italian*

Other country names make adjectives with no pattern

Bangladesh	*Bangladeshi*
Iraq	*Iraqi*
Israel	*Israeli*
Pakistan	*Pakistani*

Cyprus	*Cypriot*	Greece	*Greek*	Peru	*Peruvian*
France	*French*	Iceland	*Icelandic*	Slovakia	*Slovak*
Germany	*German*	Netherlands	*Dutch*	Thailand	*Thai*

Map punctuation

A country name in brackets shows that a place is part of that country

Corsica is part of France

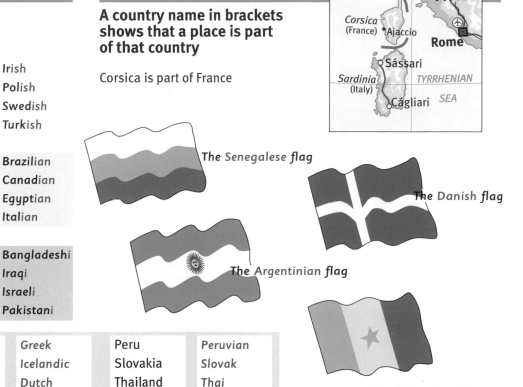

The Senegalese flag

The Danish flag

The Argentinian flag

The Bulgarian flag

6 Atlas numeracy

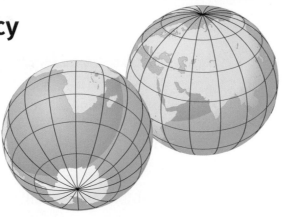

The Earth is a sphere*.

Two sets of imaginary lines help us describe where places are on the Earth.

All the lines are numbered and some have special names.

* It's actually slightly flattened at the north and south poles.

Longitude

Lines of longitude measure distance east or west of the Prime Meridian.

The **Prime Meridian** (also called the Greenwich Meridian) is at longitude 0°.

The **International Date Line** (on the other side of the Earth) is based on longitude 180°.

Latitude

Lines of latitude measure distance north or south of the equator.

The equator is at latitude 0°.

The poles are at latitude 90°N and 90°S.

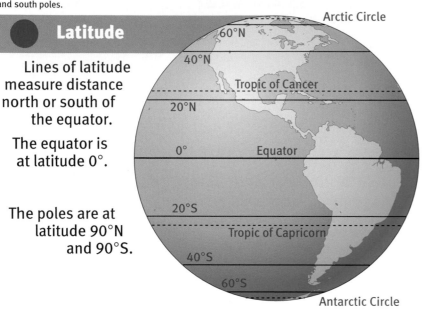

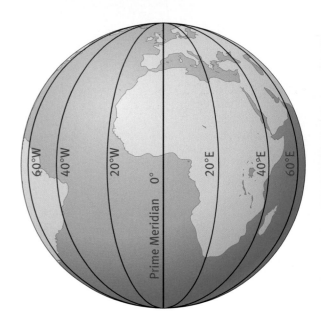

Map projections

There are many ways of showing the Earth on a flat map.

World map used in the United Kingdom and Europe

World map used in Australia and New Zealand

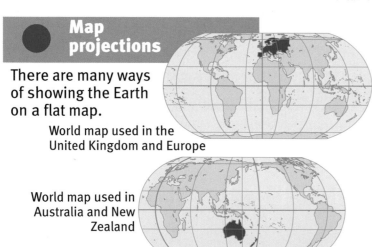

Grid codes

In this atlas, the lines of latitude and longitude are used to make a grid.

The columns of the grid have letters.

The rows of the grid have numbers.

Numbers and letters together make a grid code that can be used to describe where places are on the Earth.

Abuja is in B4   Durban is in C2

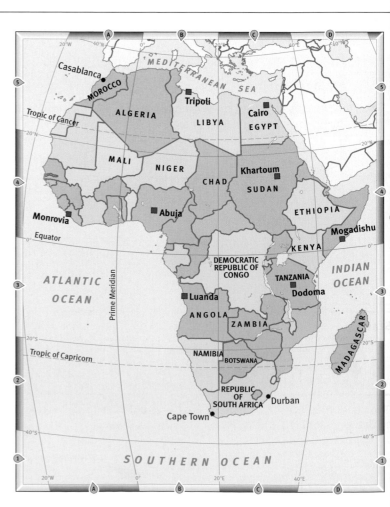

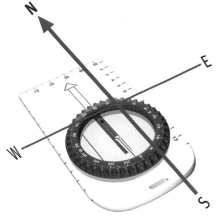

Direction

A compass is used for finding direction.

The needle of a compass always points north.

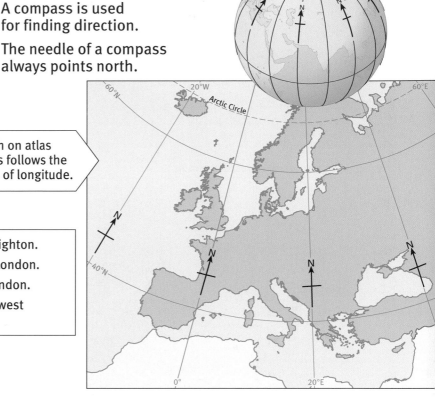

North on atlas maps follows the lines of longitude.

London is north of Brighton.

Brighton is south of London.

Reading is west of London.

Portsmouth is south west of London.

Scale

Maps are much, much smaller than the countries they show.

A few centimetres on the map stand for very many kilometres on the ground.

Each division on the scale line is one centimetre. The scale line shows how many kilometres are represented by one centimetre.

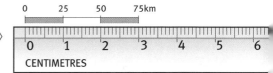

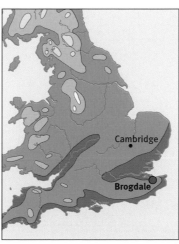

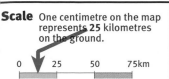

Scale One centimetre on the map represents **25 kilometres** on the ground.

The distance between Bangor and Betws-y-Coed is about 25km

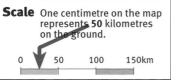

Scale One centimetre on the map represents **50 kilometres** on the ground.

The distance between Perth and Edinburgh is about 50km

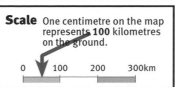

Scale One centimetre on the map represents **100 kilometres** on the ground.

The distance between Cambridge and Brogdale is about 100km

Larger scale
smaller area
more detail

Smaller scale
larger area
less detail

On world maps the scale is only true along the equator.

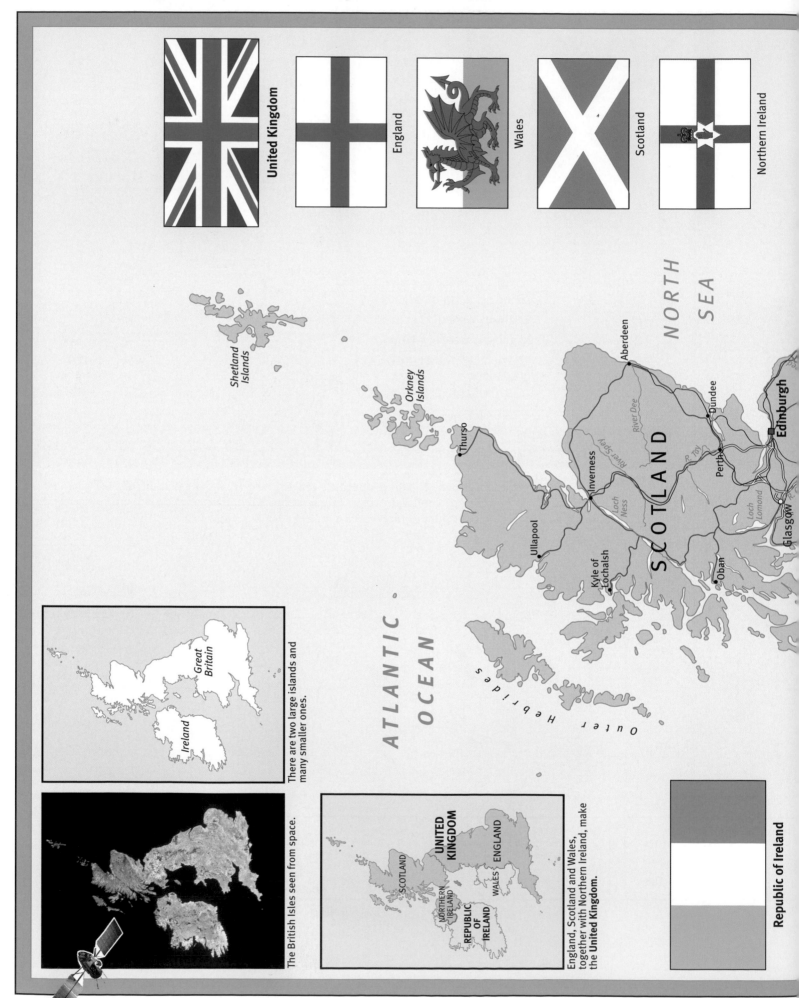

United Kingdom

England

Wales

Scotland

Northern Ireland

Republic of Ireland

Shetland Islands

Orkney Islands

Thurso

ATLANTIC OCEAN

NORTH SEA

Aberdeen

River Dee

Dundee

Edinburgh

Inverness

River Spey

Perth

R. Tay

Loch Ness

Loch Lomond

Glasgow

Ullapool

SCOTLAND

Kyle of Lochalsh

Oban

Outer Hebrides

Great Britain

Ireland

There are two large islands and many smaller ones.

The British Isles seen from space.

SCOTLAND

UNITED KINGDOM

ENGLAND

NORTHERN IRELAND

WALES

REPUBLIC OF IRELAND

England, Scotland and Wales, together with Northern Ireland, make the **United Kingdom.**

Transverse Mercator Projection

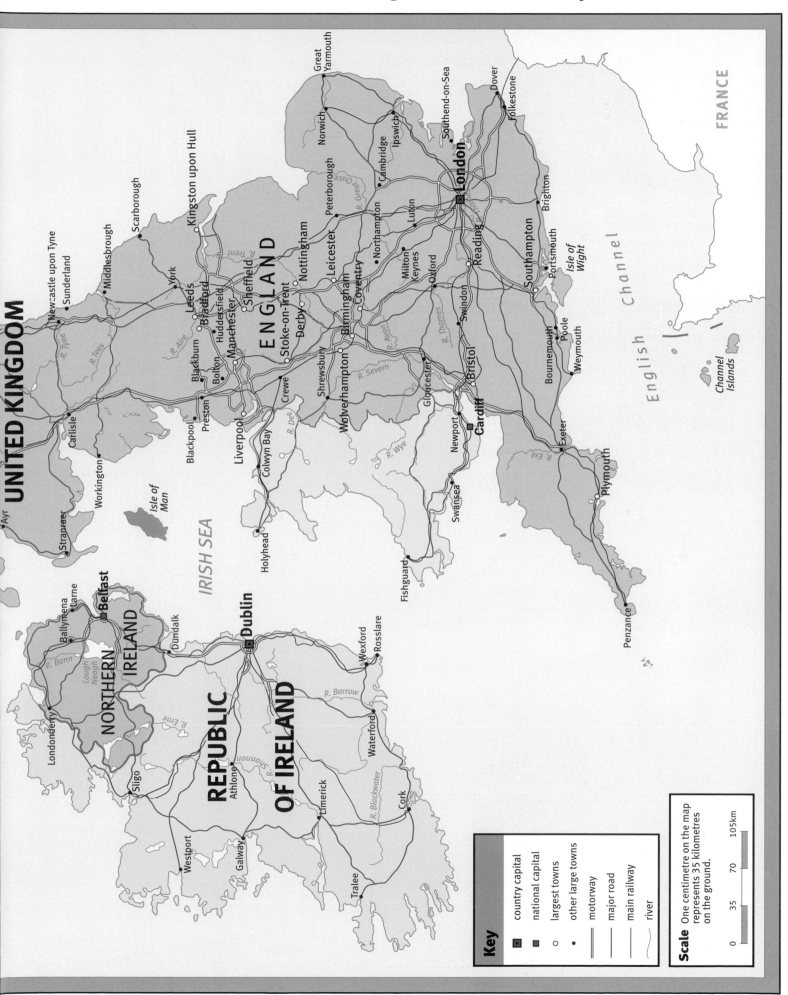

FRANCE

English Channel

Channel
Islands

IRISH SEA

UNITED KINGDOM

Ayr
Stranraer
Newcastle upon Tyne
Sunderland
Carlisle
Middlesbrough
Workington
Scarborough
R. Tyne
R. Tees
Kingston upon Hull
York
Leeds
Bradford
R. Aire
Blackburn
Huddersfield
Bolton
Manchester
Blackpool
Preston
Liverpool
Crewe
Sheffield
Stoke-on-Trent
Derby
Nottingham
R. Trent
ENGLAND
Colwyn Bay
R. Dee
Holyhead
Shrewsbury
Wolverhampton
Birmingham
Coventry
Leicester
Peterborough
R. Great Ouse
R. Severn
R. Avon
Norwich
Great
Yarmouth
Cambridge
Ipswich
Northampton
Milton
Keynes
Luton
Oxford
R. Thames
London
Southend-on-Sea
Dover
Folkestone
Reading
Swindon
Brighton
Southampton
Portsmouth
*Isle of
Wight*
Bournemouth
Poole
Weymouth
Gloucester
Bristol
Cardiff
Newport
Swansea
R. Wye
R. Exe
Exeter
Plymouth
Penzance
Fishguard
*Isle of
Man*

NORTHERN IRELAND
Belfast
Larne
Ballymena
Londonderry
R. Bann
*Lough
Neagh*
Sligo
R. Erne
Dundalk
Dublin
Westport
Galway
Athlone
Shannon
Limerick
Tralee
R. Barrow
R. Blackwater
Cork
Waterford
Wexford
Rosslare
REPUBLIC
OF IRELAND

Key

- ■ country capital
- ■ national capital
- ○ largest towns
- • other large towns
- ⎯ motorway
- ⎯ major road
- ⎯ main railway
- ⌇ river

Scale One centimetre on the map
represents 35 kilometres
on the ground.

0   35   70   105km

Key

- ▬▬▬ international boundary
- - - - national boundary
- ═══ motorway
- ─── main road
- ─── railway
- ✈ main airport
- ∿ river
- ⊥⊥⊥ canal
- ⬭ lake

towns
- ⬠ built-up area
- ⊡ largest towns
- ○ large towns
- • other towns

land height
above sea level in metres

- more than 1000m
- 500 – 1000m
- 200 – 500m
- 100 – 200m
- less than 100 metres
- land below sea level

▲ highest peaks with heights in metres

Scale One centimetre on the map represents 12.5 kilometres on the ground.

0   12.5   25   37.5km

ATLANTIC

OCEAN

Cape Wrath

Durness

Butt of Lewis

Port of Ness

Eddrachillis Bay

927m Ben Hope

Lo Loy

Broad Bay

Eye Peninsula

Lewis

Stornoway

961m Ben Klibreck

Lochinver

998m Ben More Assynt

Loch Shi

Scarp

Loch Langavat

The Minch

Enard Bay

Lairg

Clisham 799m

Taransay

Tarbert

Loch Broom

Scalpay

Shiant

Ullapool

Beinn Dearg 1081m

Bo Brid

Harris

Little Minch

Rubha Hunish

Kilmaluag

Poolewe

Gairloch

Loch Maree

Loch Torridon

1109m Sgurr Mór

Ben Wyvis 1046m

Pabbay

Berneray

Sound of Harris

Loch Snizort

Uig

Loch Fannich

Dingwal

North Uist

Lochmaddy

The Storr 719m

Sound of Raasay

Inner Sound

Conon Bridge

Muir of Ord

St. Kilda

Outer Hebrides

Benbecula

Dunvegan

Portree

Skye

Raasay

Loch Monar

Orrin Reservoir

SCOT

South Uist

Cuillin Hills

Scalpay

Kyle of Lochalsh

Carn Eige 1183m

Drumnadrochit

Loch Ness

Lochboisdale

Soay

Broadford

Invermoriston

Eriskay

Canna

Sound of Sleat

Fort Augustus

Barra

Kinloch

Calligarry

Invergarry

Castlebay

Vatersay

Rhum

Mallaig

Loch Garry

Loch Lochy

Mingulay

Eigg

Arisaig

Loch Morar

Loch Arkaig

Loch Laggan

Inner Hebrides

Muck

Sound of Arsaig

Loch Shiel

Fort William

Ben Nevis 1344m

© Oxford University Press
Transverse Mercator Projection

Fair Isle

Orkney Islands

Mull Head
Papa Westray
North Ronaldsay
Westray
Sanday
Rousay
Eday
Brough Head
Stronsay
Westray Firth
Stronsay Firth
Mainland
Shapinsay
Kirkwall
Scapa
Stromness
479m ▲ Ward Hill
Scapa Flow
Hoy
South Ronaldsay
Pentland Firth
Dunnet Head
Stroma
Duncansby Head
John o'Groats
Thurso
Halkirk
Bettyhill
River Wick
Wick
River Thurso
Kinbrace
Lybster
Loch Nan Clár
Helmsdale
Brora
Golspie
Dornoch Firth
Tarbat Ness
Tain
Invergordon
Cromarty
Moray Firth
Lossiemouth
Portknockie
Cullen
Rosehearty
Fraserburgh
Nairn
Elgin
Portsoy
Macduff
Forres
Fochabers
River Deveron
Turriff
Mintlaw
Peterhead
Rothes
Keith
Inverness
Dufftown
Huntly
A N D
Oldmeldrum
Ellen
Grantown-on-Spey
Inverurie
River Spey
River Don
onadhliath Mountains
Aviemore
Dyce
Aberdeen
Cairngorms
1244m ▲ Cairn Gorm
Kingussie
Aboyne
Newtonmore
Ballater
River Dee
Braemar
Lochnagar 1155m
Banchory
Dalwhinnie
R. North Esk
AMPIAN MOUNTAINS
Stonehaven
Inverbervie

Shetland Islands

Herma Ness
Haroldswick
Unst
Point of Fethaland
Fetlar
Yell
Yell Sound
Esha Ness
Out Skerries
St. Magnus Bay
Muckle Roe
Whalsay
Papa Stour
Mainland
Walls
Bressay
417m ▲ Foula
Scalloway
Lerwick
Shetland Islands
Sumburgh Head
Fair Isle

NORTH SEA

59°N
58°N
57°N
60°N
60°N

Scale One centimetre on the map represents 12.5 kilometres on the ground.

0   12.5   25   37.5km

GRAMPIAN MOUNTAINS

1344m Ben Nevis
Blackwater Reservoir
Kinlochleven
Loch Rannoch
Ben Lawers 1214m
Loch Lyon

Muck
Coll
Tobermory
Lochaline
Loch Shiel
Fort William
Loch Linnhe
Tiree
Ulva
Craignure
Lismore
Loch Etive
River Orchy
Tyndrum
Lochearnh
967m Ben More
Mull
Kerrera
Oban
1124m Ben Cruachan
Crianlarich
Ben More 1174m
Loch Ear
Fionnphort
Ross of Mull
Firth of Lorn
SCOTLAN
Iona
Inveraray
974m Ben Lomond
Callan
Scarba
Furnace
Strachur
Loch Lomond
River Forth
Colonsay
Scalasaig
Lochgilphead
Loch Long
Garelochhead
Campsie Fells
Oronsay
Helensburgh
Loch Fyne
Greenock
Dumbarton
Kirkintillo
Jura
Tarbert
Dunoon
Port Glasgow
Clydebank
Bearsden
Coatbri
Port Askaig
Rothesay
Firth of Clyde
Johnstone
Paisley
Glas
Craighouse
Kennacraig
Clachan
Bute
Largs
Barrhead
Hamilte
Newton Mearns
East Kilbri
Islay
Gigha
Lochranza
Goat Fell 874m
Ardrossan
Stewarton
Darvel
Portnahaven
Kilbrannan Sound
Saltcoats
Irvine
Kilmarnock
Mull of Oa
Port Ellen
Brodick
Arran
Prestwick
Ayr
River Ayr
Cumno
Sound of Bute
Campbeltown
New Cumnock
Malin Head
Rathlin Island
Mull of Kintyre
Southend
Maybole
River Doon
SOUTHER
Inishowen Peninsula
Giant's Causeway
Rathlin Sound
Fair Head
Ailsa Craig
Girvan
615m Slieve Snaght
Portrush
Bushmills
Ballycastle
Creeslough
Buncrana
Coleraine
North Channel
Ballantrae
New Galloway
Kilmacrenan
Lough Foyle
Limavady
Ballymoney
Antrim Mountains
Corsewall Point
Newton Stewart
Letterkenny
Londonderry
Dungiven
R. Bush
Cairnryan
Gatehouse of Fleet
Ballybofey
Lifford
Strabane
River Foyle
Maghera
R. Main
Carnlough
Stranraer
Glenluce
Wigtown
R. Finn
Sperrin Mountains
683m Sawel
River Bann
Ballymena
Larne
Kirkcudbrig
R. Derg
Newtownstewart
Magherafelt
Randalstown
Island Magee
Drummore
Whithorn
Donegal
NORTHERN IRELAND
Cookstown
Antrim
Carrickfergus
Mull of Galloway
Omagh
Lough Neagh
Newtownabbey
Crumlin
Belfast
Bangor
Donaghadee
Lough Derg
Coalisland
Dungannon
Lisburn
Newtownards
Point of Ay
Irvinestown
Lower Lough Erne
Aughnacloy
Portadown
Lurgan
Craigavon
Dromore
Saintfield
Ards Peninsula
Strangford Lough
Ramsey
Manorhamilton
Enniskillen
R. Blackwater
Armagh
Banbridge
R. Lagan
River Bann
Downpatrick
Kirk Michael
Snaefell 620m
R. Shannon
Monaghan
Keady
Peel
Lisnaskea
Upper Lough Erne
Newtownhamilton
Newcastle
St. John's Point
Isle of Man
Clones
Newry
852m Slieve Donard
Douglas
Castleblayney
Warrenpoint
Mourne Mtns.
Crossmaglen
Kilkeel
Port Erin
Calf of Man
Castletown

Key

— international boundary
--- national boundary
═══ motorway
—— main road
—— railway
⊕ main airport
～ river
┼┼┼ canal
◠ lake

towns

⬟ built-up area
⊡ largest towns
○ large towns
• other towns

land height
above sea level in metres

more than 1000m
500 – 1000m
200 – 500m
100 – 200m
less than 100 metres
land below sea level

▲ highest peaks with heights in metres

E F G

Pitlochry
Brechin
Milton Ness
Montrose
Kirriemuir
Forfar
Aberfeldy
Blairgowrie
Arbroath
Sidlaw Hills
Carnoustie
R. South Esk
River Tay
Dundee
Firth of Tay
Crieff
Perth
Leuchars
St. Andrews
Cupar
River Earn
Auchterarder
Auchtermuchty
nblane
Kinross
Loch Leven
Glenrothes
Anstruther
Ochil Hills
Kinross
Buckhaven
Stirling
Tillicoultry
Cowdenbeath
Kirkcaldy
North Berwick
Alloa
Dunfermline
Grangemouth
Inverkeithing
Firth of Forth
Dunbar
Falkirk
Edinburgh
Haddington
St. Abb's Head
umbernauld
Linlithgow
Musselburgh
Bathgate
Livingston
Dalkeith
Eyemouth
Airdrie
Penicuik
Lammermuir Hills
otherwell
Pentland Hills
Duns
Berwick-upon-Tweed
Wishaw
Holy Island
Lanark
Peebles
Galashiels
Coldstream
Biggar
Melrose
Bamburgh
Broad Law ▲840m
Kelso
Selkirk
Wooler
Jedburgh
The Cheviot 815m
Sanquhar
Hawick
Alnwick
Daer Reservoir
R. Aln
Moffat
Peel Fell 602m
Amble
River Coquet
Thornhill
Cheviot Hills
Ashington
Langholm
Kielder Water
R. North Tyne
R. Wansbeck
Lockerbie
Blyth
Dumfries
Cramlington
Castle Douglas
Annan
R. Irthing
Whitley Bay
Dalbeattie
Haltwhistle
Hexham
Newcastle upon Tyne
Tynemouth
South Shields
Brampton
Consett
Washington
Gateshead
Carlisle
Chester-le-Street
Sunderland
Solway Firth
Wigton
River Eden
Houghton-le-Spring
Durham
Peterlee
Cross Fell 893m
River Wear
Spennymoor
Hartlepool
Maryport
R. Derwent
Skiddaw 931m
Penrith
Bishop Auckland
Billingham
Redcar
Workington
Cockermouth
Mickle Fell 790m
Newton Aycliffe
Stockton-on-Tees
Middlesbrough
Whitehaven
Keswick
Derwent Water
Ullswater
Barnard Castle
Darlington
Thornaby-on-Tees
Guisborough
Whitby
St. Bees Head
Helvellyn ▲950m
Appleby-in-Westmorland
Brough
R. Tees
Scafell Pike 978m▲
Lake District
ENGLAND
Cleveland Hills
River Esk
Seascale
Ambleside
Richmond
North York Moors
Wast Water
Windermere
Windermere
River Swale
Northallerton
Scarborough
Coniston Water
Kendal
Leyburn
Vale of Pickering
Whernside 737m
River Ure
Great Whernside 704m
Thirsk
Pickering
Filey
Ulverston
River Lune
R. Wharfe
Ripon
Malton
Barrow-in-Furness
Ingleborough ▲723m
693m Pen-y-Ghent
Yorkshire Wolds
Bridlington
Carnforth
Settle
River Nidd
Haxby
Great Driffield
Morecambe
Lancaster
Ward's Stone 560m
Knaresborough
Harrogate
York
Heysham

NORTH SEA

PENNINES

56°N
55°N
54°N

3°W
2°W
1°W

3
2
1

H

© Oxford University Press

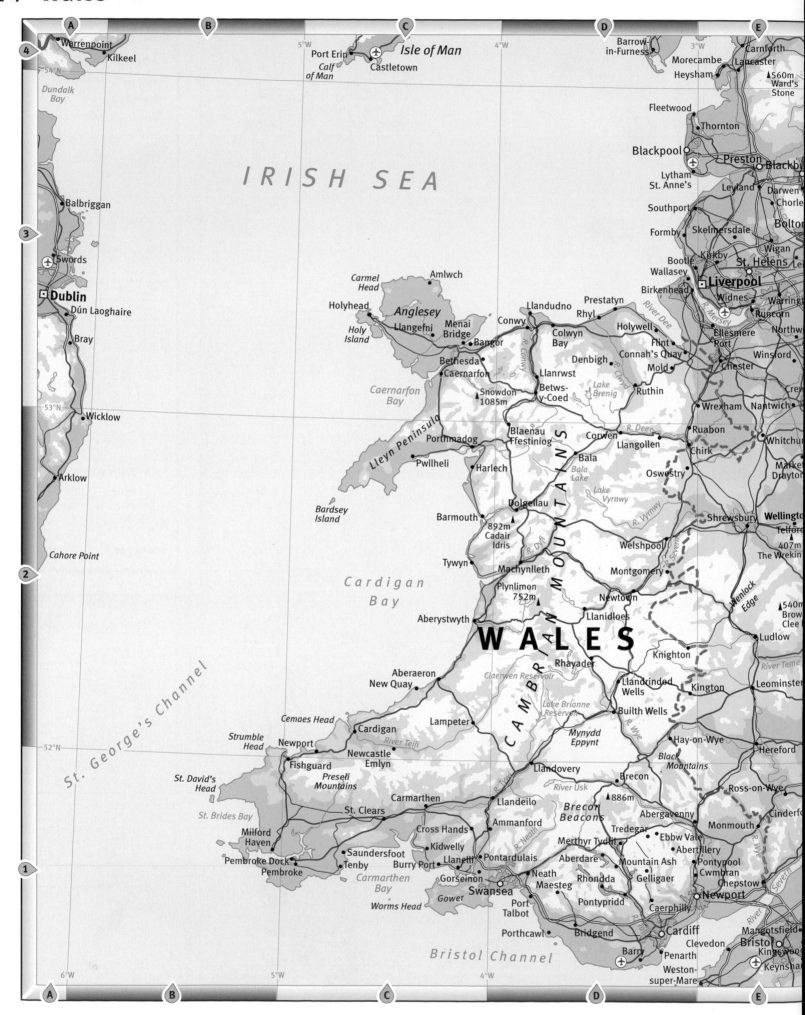

IRISH SEA

Dundalk Bay

Warrenpoint
Kilkeel
Port Erin
Calf of Man
Castletown
Isle of Man
Barrow-in-Furness
Morecambe
Lancaster
Carnforth
Heysham
▲560m Ward's Stone
Fleetwood
Thornton
Blackpool
Preston
Blackburn
Lytham St. Anne's
Leyland
Darwen
Chorley
Southport
Formby
Skelmersdale
Bolton
Wigan
St. Helens
Lee
Bootle
Kirkby
Wallasey
Birkenhead
Liverpool
Widnes
Warrington
Runcorn
Ellesmere Port
Northw
Balbriggan
Swords
Dublin
Dún Laoghaire
Bray
Wicklow
Arklow
Cahore Point

Carmel Head
Amlwch
Holyhead
Anglesey
Holy Island
Llangefni
Menai Bridge
Bangor
Bethesda
Caernarfon
Caernarfon Bay
Snowdon 1085m
Conwy
Llandudno
Rhyl
Prestatyn
Holywell
Colwyn Bay
Connah's Quay
Denbigh
Mold
Chester
Winsford
Crew
Llanrwst
Betws-y-Coed
Lake Brenig
Ruthin
Wrexham
Nantwich
Lleyn Peninsula
Porthmadog
Blaenau Ffestiniog
Corwen
R. Dee
Llangollen
Ruabon
Whitchu
Pwllheli
Harlech
Bala
Bala Lake
Oswestry
Chirk
Market Drayton
Lake Vyrnwy
Bardsey Island
Dolgellau
R. Vyrnwy
Barmouth
892m Cadair Idris
Shrewsbury
Wellington
Welshpool
R. Severn
Telfo
▲407m The Wrekin
Tywyn
Machynlleth
Montgomery
Plynlimon 752m
Newtown
Wenlock Edge
▲540m Brow Clee
Cardigan Bay
Aberystwyth
Llanidloes
WALES
Ludlow
Rhayader
Knighton
River Teme
Leominster
Aberaeron
New Quay
Claerwen Reservoir
Llandrindod Wells
Kington
Lake Brianne Reservoir
Builth Wells
Cemaes Head
Cardigan
River Teifi
Lampeter
Mynydd Eppynt
R. Wye
Hay-on-Wye
Hereford
Strumble Head
Newport
Newcastle Emlyn
St. George's Channel
St. David's Head
Fishguard
Preseli Mountains
Llandovery
River Usk
Brecon
Black Mountains
Ross-on-Wye
Carmarthen
Llandeilo
▲886m
Brecon Beacons
Abergavenny
Monmouth
St. Brides Bay
St. Clears
Ammanford
Cross Hands
R. Neath
Merthyr Tydfil
Tredegar
Ebbw Vale
Abertillery
Cinderfo
Milford Haven
Kidwelly
Aberdare
Mountain Ash
Pontypool
Cwmbran
Pembroke Dock
Saundersfoot
Tenby
Llanelli
Burry Port
Pontardulais
Neath
Rhondda
Gelligaer
Chepstow
Pembroke
Gorseinon
Maesteg
Pontypridd
Caerphilly
Newport
Carmarthen Bay
Gower
Swansea
Port Talbot
Caerphilly
Mangotsfield
Worms Head
Porthcawl
Bridgend
Cardiff
Clevedon
Bristol
Barry
Penarth
Kingswo
Bristol Channel
Weston-super-Mare
Keynsha

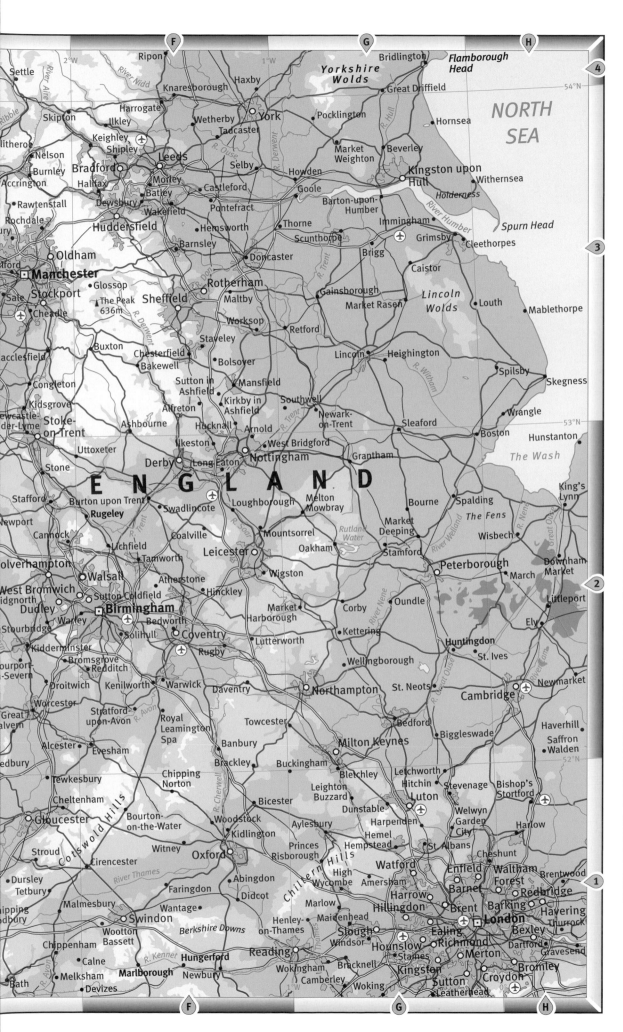

Key

▬▬▬	international boundary
▬ ▬ ▬	national boundary
▭▭▭	motorway
▬▬▬	main road
▬▬▬	railway
⊕	main airport
∼∼	river
⊢⊢⊢	canal
◡	lake

towns

▲	built-up area
⊡	largest towns
○	large towns
•	other towns

land height

above sea level in metres

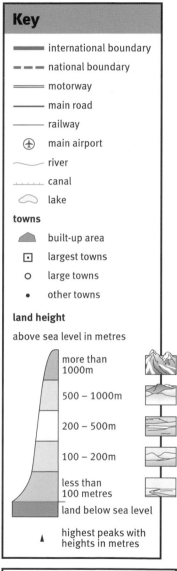

more than 1000m	
500 – 1000m	
200 – 500m	
100 – 200m	
less than 100 metres	
land below sea level	
▲	highest peaks with heights in metres

Scale One centimetre on the map represents 12.5 kilometres on the ground.

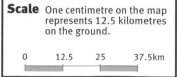

0   12.5   25   37.5km

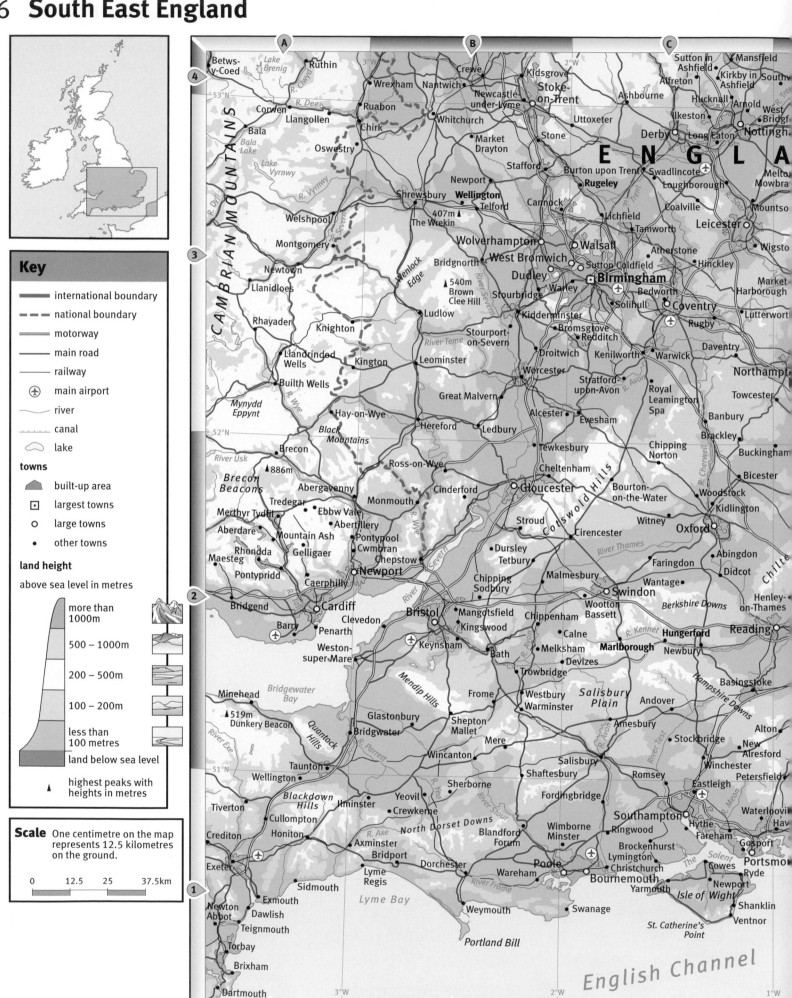

Key

▬▬▬	international boundary
▬ ▬ ▬	national boundary
══════	motorway
———	main road
———	railway
✈	main airport
～～～	river
⋯⋯⋯	canal
◠	lake

towns

⬟	built-up area
⊡	largest towns
○	large towns
•	other towns

land height

above sea level in metres

- more than 1000m
- 500 – 1000m
- 200 – 500m
- 100 – 200m
- less than 100 metres
- land below sea level
- ▲ highest peaks with heights in metres

Scale One centimetre on the map represents 12.5 kilometres on the ground.

0   12.5   25   37.5km

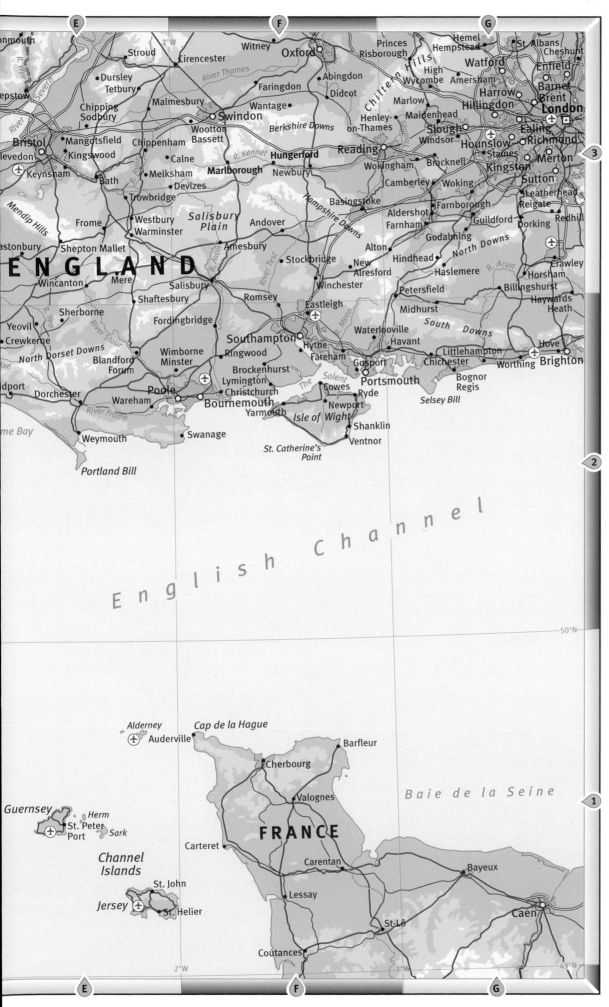

Key

▬▬▬	international boundary
▬ ▬ ▬	national boundary
═══	motorway
▬▬▬	main road
───	railway
⊕	main airport
∿	river
┈┼┈	canal
◠	lake

towns

◣	built-up area
⊡	largest towns
○	large towns
•	other towns

land height

above sea level in metres

more than 1000m	
500 – 1000m	
200 – 500m	
100 – 200m	
less than 100 metres	
land below sea level	
▲	highest peaks with heights in metres

Scale One centimetre on the map represents 12.5 kilometres on the ground.

0   12.5   25   37.5km

Highest mountains

few places in Britain are more than 1000 metres high

Mountains

steep rocky places

Moors and uplands

high windswept places with heather and rough grass

Hills

smooth slopes and gentle valleys

Low land

flat marshy land with wide rivers

© Oxford University Press

Key

colours show land height above sea level in metres

more than 1000m

500 – 1000m

200 – 500m

100 – 200m

less than 100 metres

land below sea level

▲ highest peaks with heights in metres

river

lake

Scale One centimetre on the map represents 45 kilometres on the ground.

0   45   90   135km

Shetland Islands

Orkney Islands

Cape Wrath

Outer Hebrides

Lewis

Skye

NORTHWEST HIGHLANDS

Great Glen

Loch Ness

River Spey

Cairngorms

River Dee

1344m ▲ Ben Nevis

GRAMPIAN MOUNTAINS

Mull

R. Tay

Islay

Loch Lomond

Firth of Forth

NORTH SEA

R. Clyde

Firth of Clyde

SOUTHERN UPLANDS

River Tweed

Cheviot Hills

North Channel

R. Tyne

Antrim Mountains

R. Bann

Loch Neagh

River Erne

Firth of Clyde

Lake District

River Eden

River Tees

978m ▲ Scafell Pike

North York Moors

▲852m Slieve Donard

Isle of Man

IRISH SEA

PENNINES

River Ouse

River Aire

Loch Corrib

River Shannon

River Boyne

River Liffey

Ireland

Wicklow Mountains

River Humber

R. Mersey

Anglesey

1085m ▲ Snowdon

Great Britain

River Trent

The Wash

R. Wensum

R. Barrow

River Dee

CAMBRIAN MOUNTAINS

River Severn

River Avon

The Fens

River Great Ouse

River Suir

River Blackwater

River Teifi

River Wye

River Stour

▲1041m Carrantuohill

St. George's Channel

Cardigan Bay

R. Tywi

River Usk

Brecon Beacons

Cotswold Hills

Chiltern Hills

River Thames

Bristol Channel

Salisbury Plain

North Downs

ATLANTIC OCEAN

Exmoor

South Downs

R. Exe

Dartmoor

Isle of Wight

Strait of Dover

Land's End

English Channel

Isles of Scilly

Channel Islands

Winter

| Jan. | Feb. | Mar. | Apr. | May | June | July | Aug. | Sep. | Oct. | Nov. | Dec. |

▲

average temperature

16°C
14°C
12°C
10°C
8°C
6°C
4°C
2°C
0°C
-2°C

Summer

| Jan. | Feb. | Mar. | Apr. | May | June | July | Aug. | Sep. | Oct. | Nov. | Dec. |

▲

average temperature

16°C
14°C
12°C
10°C
8°C
6°C
4°C
2°C
0°C
-2°C

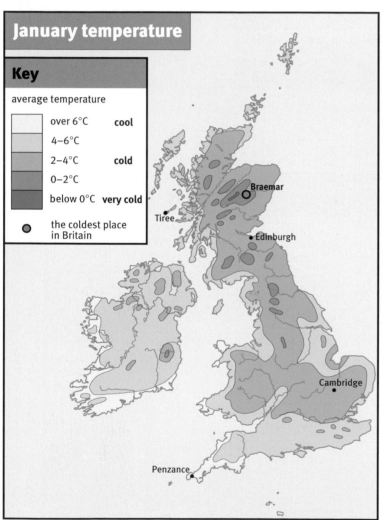

January temperature

Key

average temperature

	over 6°C	**cool**
	4–6°C	
	2–4°C	**cold**
	0–2°C	
	below 0°C	**very cold**

⊙ the coldest place in Britain

Braemar
Tiree
Edinburgh
Cambridge
Penzance

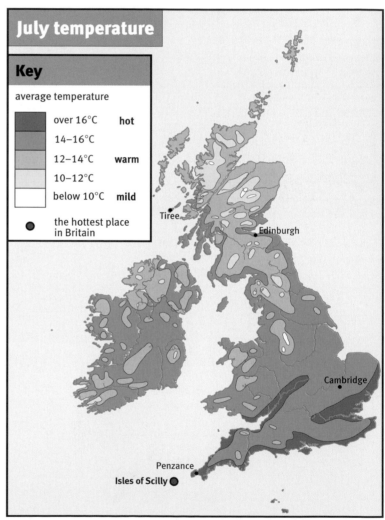

July temperature

Key

average temperature

	over 16°C	**hot**
	14–16°C	
	12–14°C	**warm**
	10–12°C	
	below 10°C	**mild**

● the hottest place in Britain

Tiree
Edinburgh
Cambridge
Penzance
Isles of Scilly ●

Climate regions

Tiree
Edinburgh
Cambridge
Penzance

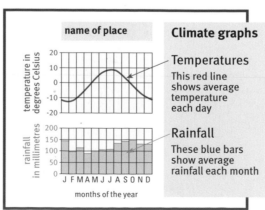

name of place

Climate graphs

temperature in degrees Celsius

20
10
0
-10
-20

Temperatures

This red line shows average temperature each day

rainfall in millimetres

200
150
100
50

J F M A M J J A S O N D

months of the year

Rainfall

These blue bars show average rainfall each month

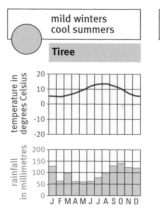

mild winters cool summers

Tiree

temperature in degrees Celsius

20
10
0
-10
-20

rainfall in millimetres

200
150
100
50

J F M A M J J A S O N D

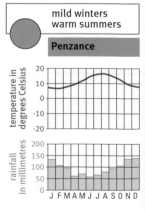

mild winters warm summers

Penzance

temperature in degrees Celsius

20
10
0
-10
-20

rainfall in millimetres

200
150
100
50

J F M A M J J A S O N D

average rainfall

2400mm — very wet
2200mm — quite wet
2000mm
1800mm
1600mm — wet
1400mm
1200mm
1000mm — quite dry
800mm
600mm — very dry
400mm
200mm

| Jan. | Feb. | Mar. | Apr. | May | June | July | Aug. | Sep. | Oct. | Nov. | Dec. |

Wet winds from the west rise and cool to give rain and snow. Mountains are the wettest places in the British Isles.

West is wetter

East is drier

Annual rainfall

Key

average rainfall in a year

- more than 2400mm
- 1500–2400mm
- 800–1500mm
- 600–800mm
- less than 600mm
- ● the wettest place in Britain
- ○ the driest place in Britain

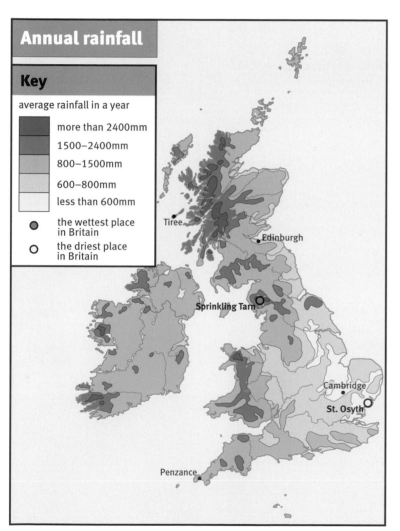

Tiree

Edinburgh

Sprinkling Tarn

Cambridge

St. Osyth

Penzance

Water supply

Key

- high land (above 200m)
- low land (below 200m)
- ● very large reservoirs (holding more than 50 million cubic metres of water*)
- built-up area

*You could fit more than 100 000 houses into these reservoirs

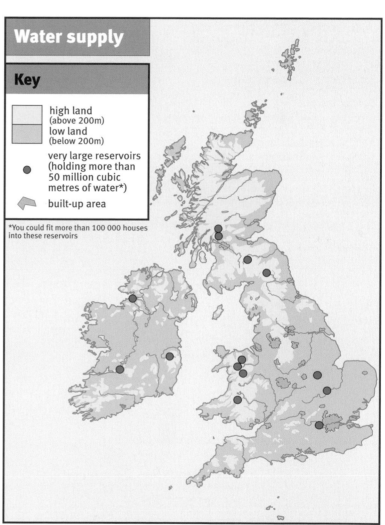

cold winters cool summers

Edinburgh

cool winters warm summers

Cambridge

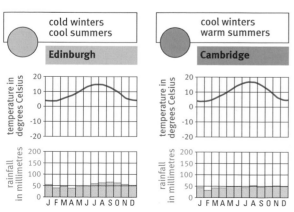

temperature in degrees Celsius
20
10
0
-10
-20

rainfall in millimetres
200
150
100
50

J F M A M J J A S O N D

temperature in degrees Celsius
20
10
0
-10
-20

rainfall in millimetres
200
150
100
50

J F M A M J J A S O N D

The water cycle

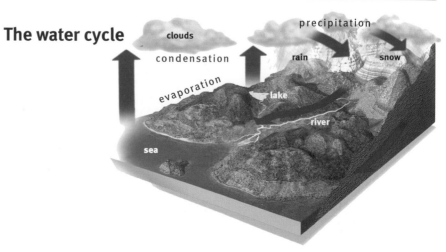

clouds

precipitation

condensation

rain

snow

evaporation

lake

river

sea

Cities and towns

People live in settlements of different sizes

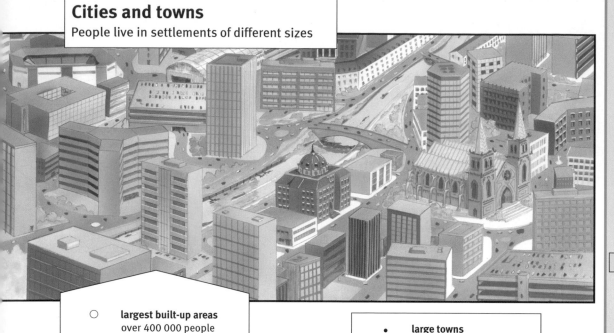

○ **largest built-up areas**
over 400 000 people

• **large towns**
25 000 – 100 000

◉ **largest towns**
100 000 – 400 000 people

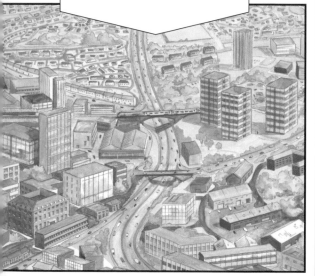

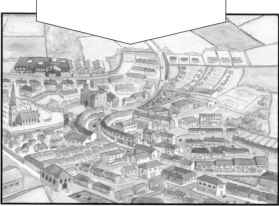

small towns and villages
under 25 000 people
(not shown on the map)

Population density

The number of people that live in an area

very crowded
over 250 people living
in a square kilometre

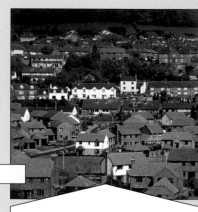

quite crowded
50 – 250 people living
in a square kilometre

quite empty
under 50 people living
in a square kilometre

Where people live

If there were 100 people in the
United Kingdom, this is where
they would live:

† England
† Scotland
† Wales
† Northern
  Ireland

Population pyramid

If there were 100 people in the United Kingdom,
this is how old they would be:

90 years old and over	
80 years old and over	
between 70 and 79	
between 60 and 69	
between 50 and 59	
between 40 and 49	
between 30 and 39	
between 20 and 29	
between 10 and 19	
9 years old and under	

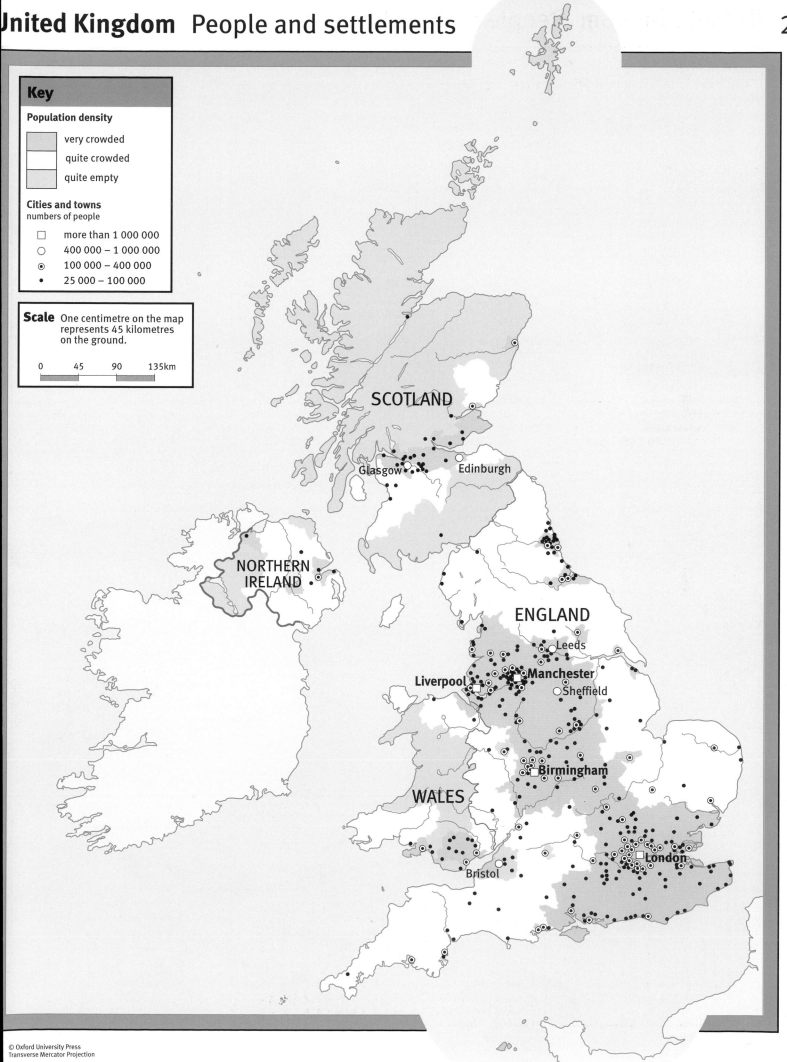

Key

Population density

very crowded

quite crowded

quite empty

Cities and towns
numbers of people

☐ more than 1 000 000

○ 400 000 – 1 000 000

◉ 100 000 – 400 000

• 25 000 – 100 000

Scale One centimetre on the map
represents 45 kilometres
on the ground.

0   45   90   135km

SCOTLAND

Glasgow   Edinburgh

NORTHERN
IRELAND

ENGLAND

Leeds

Liverpool   Manchester

Sheffield

WALES   Birmingham

Bristol   London

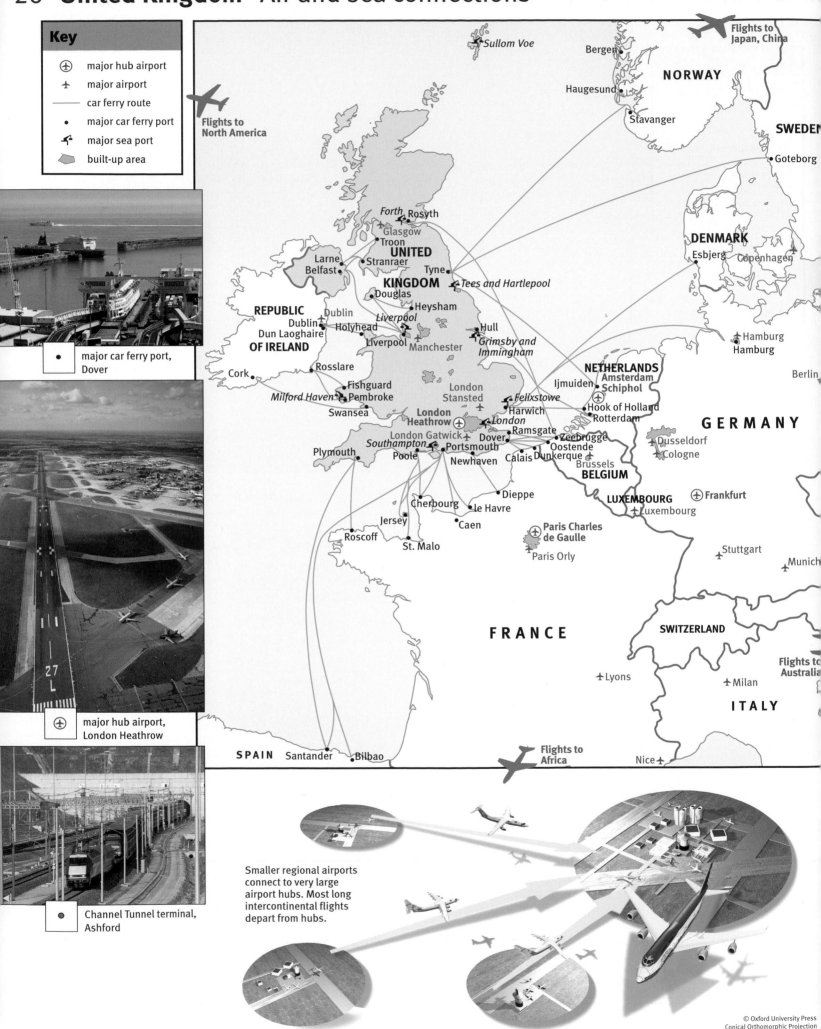

Key

- ⊕ major hub airport
- ✈ major airport
- — car ferry route
- • major car ferry port
- ⚓ major sea port
- ▨ built-up area

• major car ferry port, Dover

⊕ major hub airport, London Heathrow

• Channel Tunnel terminal, Ashford

NORWAY

Flights to Japan, China

Sullom Voe
Bergen
Haugesund
Stavanger

SWEDEN

Goteborg

DENMARK

Esbjerg
Copenhagen

Flights to North America

UNITED KINGDOM

Forth Rosyth
Glasgow
Troon
Larne Stranraer
Belfast Tyne
Tees and Hartlepool
Douglas
Heysham
Dublin Liverpool Hull
Dublin Holyhead Grimsby and
Dun Laoghaire Liverpool Immingham
Manchester

REPUBLIC OF IRELAND

Cork
Rosslare
Fishguard London
Milford Haven Pembroke Stansted
Swansea Felixstowe
Harwich Hook of Holland
London Ramsgate Rotterdam
Heathrow London
London Gatwick Dover
Plymouth Southampton Portsmouth Zeebrugge
Poole Newhaven Calais Oostende
Dieppe Dunkerque Brussels

Hamburg
Hamburg
Berlin

NETHERLANDS
Amsterdam
Schiphol
Ijmuiden

GERMANY

Dusseldorf
Cologne

BELGIUM

LUXEMBOURG
⊕ Frankfurt
Luxembourg

Cherbourg le Havre
Jersey Caen
Roscoff
Plymouth St. Malo Paris Charles
de Gaulle
Paris Orly

Stuttgart
Munich

FRANCE

SWITZERLAND

Lyons
Milan

ITALY

SPAIN Santander Bilbao

Flights to Africa Nice

Flights to Australia

Smaller regional airports connect to very large airport hubs. Most long intercontinental flights depart from hubs.

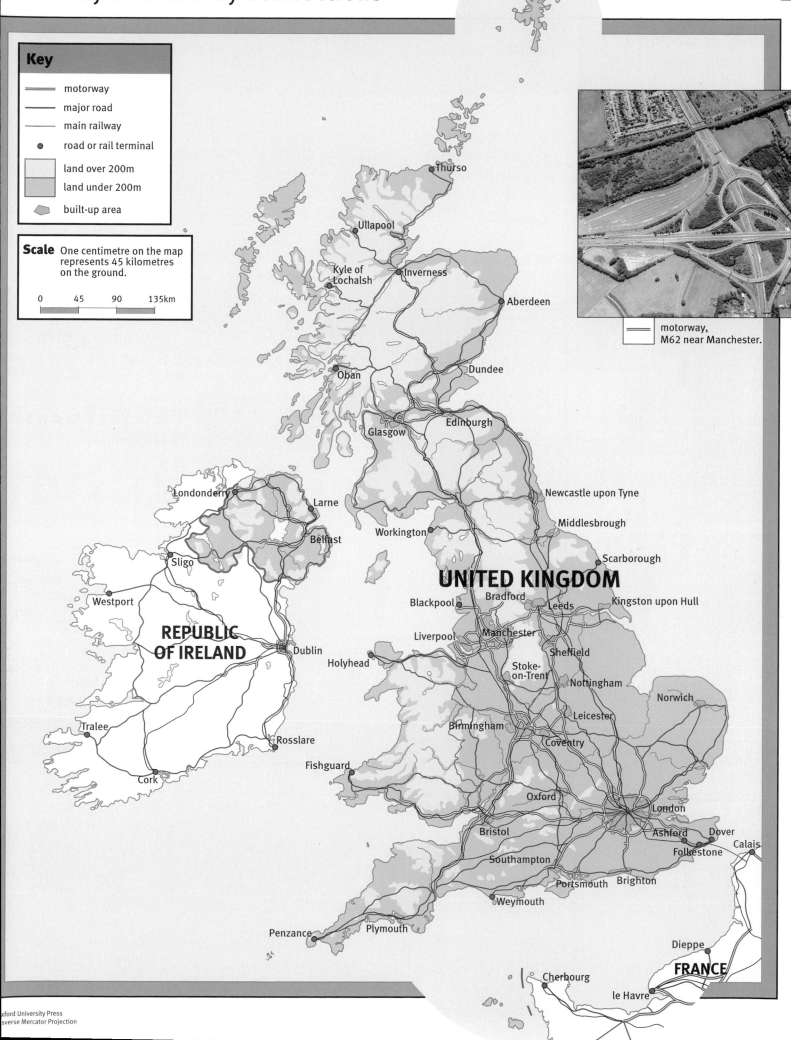

Key

═══	motorway
───	major road
───	main railway
●	road or rail terminal
	land over 200m
	land under 200m
	built-up area

Scale One centimetre on the map represents 45 kilometres on the ground.

0   45   90   135km

motorway, M62 near Manchester.

Thurso

Ullapool

Kyle of Lochalsh

Inverness

Aberdeen

Oban

Dundee

Edinburgh

Glasgow

Londonderry

Larne

Belfast

Sligo

Workington

Newcastle upon Tyne

Middlesbrough

Scarborough

Westport

REPUBLIC OF IRELAND

UNITED KINGDOM

Blackpool   Bradford   Kingston upon Hull

Leeds

Liverpool   Manchester

Dublin

Holyhead

Sheffield

Stoke-on-Trent

Nottingham

Norwich

Leicester

Tralee

Rosslare

Birmingham

Coventry

Fishguard

Oxford

London

Cork

Bristol

Ashford   Dover

Folkestone   Calais

Southampton

Portsmouth   Brighton

Weymouth

FRANCE

Penzance   Plymouth

Dieppe

Cherbourg

le Havre

xford University Press
sverse Mercator Projection

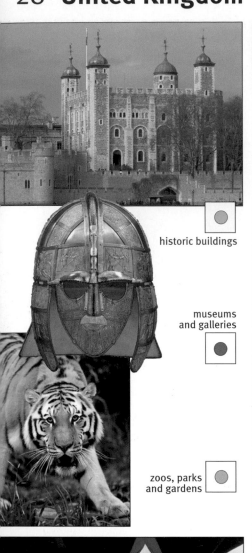

Top UK tourist attractions

Key

Symbol **colour** shows the type of tourist attraction

- historic buildings
- museums and galleries
- zoos, parks and gardens
- theme parks and piers

Symbol **size** shows how popular the attraction is

- ○ over 4 million visitors each year
- ○ 2–4 million visitors each year
- ○ 1–2 million visitors each year
- built-up area

historic buildings

museums and galleries

zoos, parks and gardens

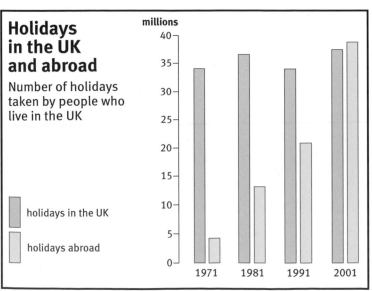

theme parks and piers

Drumpellier Country Park
Edinburgh Castle
Kelvingrove Art Gallery and Museum
Strathclyde Country Park
Windermere Lake Cruises
Flamingoland Theme Park and Zoo, Kirby Misperton
Blackpool Pleasure Beach
York Minster
Pleasureland Theme Park Southport
Upper Derwent Reservoirs
Chester Zoo
Alton Towers
Drayton Manor Family Theme Park, Tamworth
Pleasure Beach, Great Yarmouth
Fairlands Valley Park
Ashton Court Estate
Legoland, Windsor
Kew Gardens
Canterbury Cathedral
Eden Project
Eastbourne Pier

Inner London

Madam Tussaud's
British Museum
National Gallery
National Portrait Gallery
London Eye
Science Museum
Tower of London
Natural History Museum
Tate Modern
Tate Britain
Victoria & Albert Museum
Westminster Abbey

Holidays in the UK and abroad

Number of holidays taken by people who live in the UK

millions

- holidays in the UK
- holidays abroad

	1971	1981	1991	2001

Holidays abroad

Canada
USA

each symbol stands for 1 million British tourists

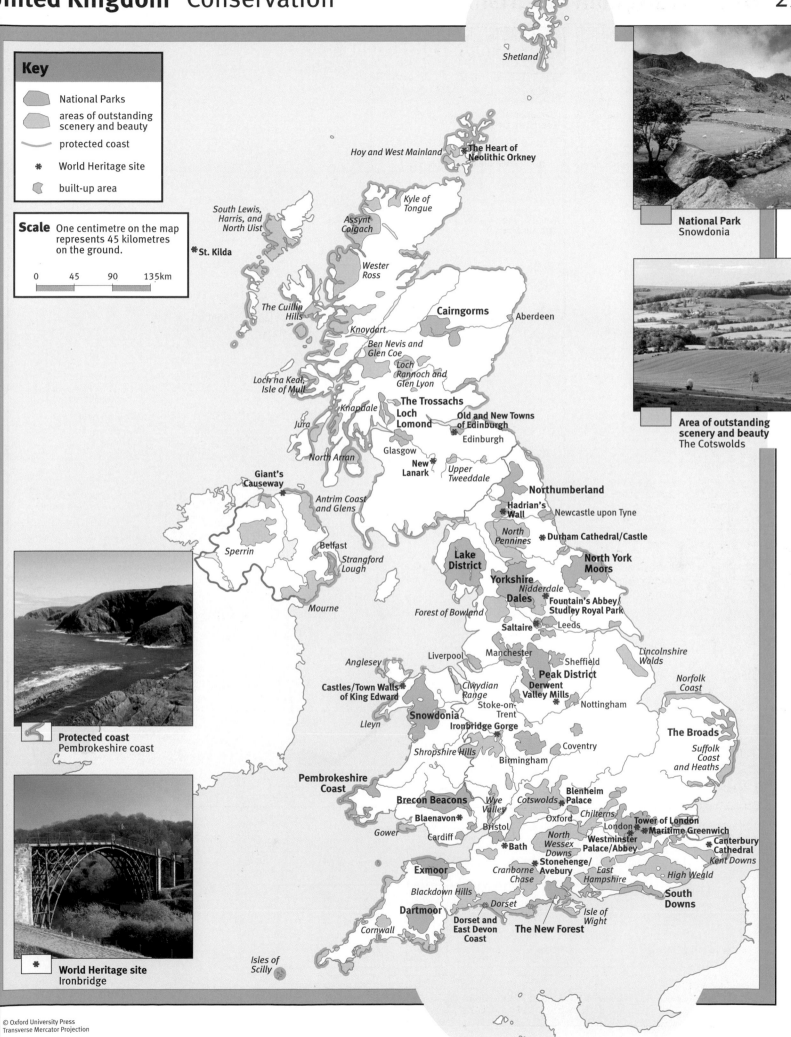

Key

- National Parks
- areas of outstanding scenery and beauty
- protected coast
- ✳ World Heritage site
- built-up area

Scale One centimetre on the map represents 45 kilometres on the ground.

0   45   90   135km

National Park
Snowdonia

Area of outstanding scenery and beauty
The Cotswolds

Protected coast
Pembrokeshire coast

World Heritage site
Ironbridge

Shetland

Hoy and West Mainland   ✳ **The Heart of Neolithic Orkney**

Kyle of Tongue

South Lewis, Harris, and North Uist

Assynt Coigach

✳ **St. Kilda**

Wester Ross

The Cuillin Hills

Knoydart

Cairngorms   Aberdeen

Ben Nevis and Glen Coe

Loch Rannoch and Glen Lyon

Loch na Keal, Isle of Mull

Knapdale

Jura

The Trossachs Loch Lomond

Old and New Towns of Edinburgh

Edinburgh

North Arran

Glasgow

New Lanark ✳

Upper Tweeddale

Giant's Causeway ✳

Antrim Coast and Glens

Northumberland

Hadrian's Wall ✳

Newcastle upon Tyne

Belfast

Sperrin

Strangford Lough

North Pennines

✳ **Durham Cathedral/Castle**

Lake District

North York Moors

Mourne

Yorkshire Dales

Nidderdale

✳ **Fountain's Abbey/ Studley Royal Park**

Forest of Bowland

Saltaire ✳   Leeds

Anglesey

Liverpool   Manchester

Sheffield

Lincolnshire Wolds

Castles/Town Walls of King Edward ✳

Clwydian Range

Peak District

Derwent Valley Mills ✳

Norfolk Coast

Stoke-on-Trent   Nottingham

Snowdonia

Lleyn

Ironbridge Gorge

Shropshire Hills

Coventry

Birmingham

The Broads

Suffolk Coast and Heaths

Pembrokeshire Coast

Brecon Beacons

Wye Valley

Cotswolds   **Blenheim Palace** ✳

Blaenavon ✳

Oxford   Chilterns

Gower   Cardiff   Bristol

North Wessex Downs

Westminster Palace/Abbey ✳

London ✳   **Tower of London** ✳

✳ **Maritime Greenwich**

✳ **Canterbury Cathedral**

Kent Downs

✳ **Bath**

Stonehenge/ Avebury ✳

Cranborne Chase

East Hampshire

High Weald

South Downs

Exmoor

Blackdown Hills

✳ Dorset   Isle of Wight

Dartmoor

Cornwall

Dorset and East Devon Coast

The New Forest

Isles of Scilly

Key

land height in metres above sea level

more than 2000m

1000 – 2000m

500 – 1000m

200 – 500m

less than 200 metres

land below sea level

▲ highest peaks with heights in metres

lake

river

Scale One centimetre on the map represents 240 kilometres on the ground.

0   240   480   720km

© Oxford University Press
Conical Orthomorphic Projection

A B C D E F

3

2

1

20°W   0°   20°E   40°E   60°E

Arctic Circle

60°N

Prime Meridian

ICELAND
Reykjavik

N

ATLANTIC
OCEAN

20°W

**RUSSIAN
FEDERATION
(RUSSIA)**

60°N

FINLAND

Oslo   Helsinki   St. Petersburg
Nizhniy-
Novgorod

Stockholm   Tallinn
ESTONIA

Belfast   Edinburgh   NORTH
SEA   Göteborg   **LATVIA**
Riga   Moscow

**REPUBLIC
OF IRELAND**   **UNITED**
Dublin   **KINGDOM**   Manchester   **DENMARK**
Copenhagen   **LITHUANIA**
Vilnius   Minsk

Birmingham   KALININGRAD
(Russia)   **BELARUS**

London   Rotterdam   NETHERLANDS   Hamburg   **POLAND**   Warsaw   Kiev   Kharkov   Volgograd
Amsterdam   Berlin
GERMANY   Düsseldorf   **UKRAINE**   Donets'k   Rostov-on-Don
BELGIUM
Brussels   LUXEMBOURG
Paris   Luxembourg   Prague   Krakow
CZECH REP.   **SLOVAKIA**   Chisinau   MOLDOVA   Odessa
FRANCE   Bern   Munich   Vienna   Bratislava
Bordeaux   SWITZERLAND   LIECHTENSTEIN   **AUSTRIA**   Budapest
Lyons   Ljubljana   **HUNGARY**   **ROMANIA**   BLACK   SEA   GEORGIA
Oporto   Milan   SLOVENIA   Zagreb   Bucharest   T'bilisi
PORTUGAL   **SPAIN**   ANDORRA   Marseilles   MONACO   CROATIA   BOSNIA-   Belgrade
Lisbon   Madrid   Barcelona   SAN   HERZEGOVINA
MARINO   Sarajevo   SERBIA AND   **BULGARIA**
Valencia   Rome   MONTENEGRO   Sofia   Istanbul
Gibraltar   Seville   Naples   Tiranë   Skopje   Ankara
(UK)   FYRO
Ceuta   Melilla   MACEDONIA   **TURKEY**
(Sp.)   (Sp.)   ALBANIA   **GREECE**   Izmir   Adana

MOROCCO   MEDITERRANEAN   Athens

40°N

0°

ITALY

20°E

Valletta   Nicosia   SYRIA
MALTA   **CYPRUS**   LEBANON   IRAQ

TUNISIA   SEA   ISRAEL
GAZA   JORDAN

LIBYA   EGYPT   SAUDI
ARABIA

Tropic of Cancer

20°E   40°E

Key

colours show
countries

ITALY   country names are
labelled like this

capital cities

other important cities

© Oxford University Press

32 Europe

Key

——	country boundary
– – –	disputed boundary
——	motorway or main road
—	railway
✈	main airport
	river
	lake

towns and cities

■	capital cities
○	largest towns
•	other large towns

land height

above sea level in metres

more than 5000m	
2000 – 5000m	
1000 – 2000m	
500 – 1000m	
200 – 500m	
less than 200 metres	
land below sea level	
▲	highest peaks with heights in metres

Scale One centimetre on the map represents 150 kilometres on the ground.

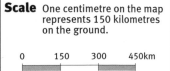

0   150   300   450km

ATLANTIC OCEAN

20°W   60°N   10°W   0°   10°E

Shetland Islands

Outer Hebrides

Orkney Islands

Bergen

NORWAY

Oslo

Lake Vänern

Lake Vätter

Göteb

Inverness

1344m Ben Nevis ▲

Aberdeen

Glasgow   Dundee   Edinburgh

NORTH SEA

DENMARK

Copenhagen   Ma

Bornho

Galway

Belfast

REPUBLIC OF IRELAND   Dublin   Manchester   UNITED KINGDOM

Cork   Birmingham

Cardiff

NETHERLANDS

Frisian Is.

Hamburg   Szczec

R. Elbe

The Hague   Amsterdam

Rotterdam

Hannover   Be

London

BELGIUM

Düsseldorf

GERMANY

Brussels

LUXEMBOURG   Prague

Luxembourg   CZE

Brest

Paris   Nürnberg

Strasbourg   R. Danube

R. Rhine

Bay of Biscay

Nantes   FRANCE   Munich

R. Loire   R. Seine

AUSTRIA

A Coruña   Bern   LIECHTENSTEIN

Cape Finisterre   Lyons   1807m Mont Blanc ▲   SWITZERLAND   Ljubljana

Bordeaux   A L P S   Milan   SLO

MASSIF CENTRAL   Verona   Zag

Bilbao   R. Rhône   Turin   R. Po

Oporto   Cantabrian Mts.   Toulouse   SAN MARINO

R. Douro   Zaragoza   PYRÉNÉES   Marseilles   MONACO   Florence

PORTUGAL   R. Duero   ANDORRA   ITALY   APPENNINES

R. Ebro   Corsica (France)

SPAIN   Barcelona   Ajaccio   Rome

Lisbon   Madrid

R. Tagus   Valencia   Balearic Islands   Menorca   Sardinia (Italy)   Sássari   Naples

Faro   Seville   R. Guadalquivir   Ibiza   Mallorca   Cágliari   TYRRHENIAN SEA

Cape St. Vincent   Cádiz

Tangier   Gibraltar (UK)   MEDITERRANEAN   Réggio di Calabria

Ceuta (Sp.)   Palermo

Rabat   Melilla (Sp.)   Oran   Algiers   Annaba   Mt Etna 3323m   Sicily

Casablanca   MOROCCO   Tunis

TUNISIA   Valle

Sfax   MALTA

ATLAS   MOUNTAINS

Béchar   30°N

A L G E R I A   Tripoli

0°   10°E   LIBYA

© Oxford University Press

Conical Orthomorphic Projection

FINLAND

Helsinki
Tallinn
ESTONIA
Stockholm
Gotland
G. of Riga
LATVIA
Riga
LITHUANIA
Kaliningrad RUSSIA
Vilnius
Gdansk
BELARUS
Minsk
North European Plain
POLAND
Warsaw
Wroclaw
Krakow
L'viv
EP.
SLOVAKIA
Bratislava
CARPATHIANS
Vienna
Budapest
HUNGARY
ROMANIA
MOLDOVA
Chisinau
BOSNIA–HERZEGOVINA
Belgrade
Bucharest
Sarajevo
R. Danube
Constanta
SERBIA AND MONTENEGRO
Sofia
BULGARIA
Skopje
FYRO MACEDONIA
Tiranë
ALBANIA
Taranto
Thessaloníki
Mt. Olympus 2917m
GREECE
PINDHOS Mts.
IONIAN SEA
Athens
Peloponnese
SEA

Lake Onega
Lake Ladoga
Vologda
St. Petersburg
Lake Peipus
Nizhniy-Novgorod Kazan
R. Volga
Rybinsk Reservoir
Moscow
RUSSIAN FEDERATION (RUSSIA)
Samara
R. Volga
Kiev
Kharkov
Volgograd
UKRAINE
Dnipropetrovsk
Donets'k
Rostov-on-Don
R. Don
R. Dnieper
R. Dniester
Odessa
SEA OF AZOV
Crimea
Mt. Elbrus 5642m
Sevastopol
CAUCASUS MTS
GEORGIA
BLACK SEA
Samsun
Istanbul
Sivas
Ankara
TURKEY
Kayseri
Bursa
Konya
Adana
Taurus Mountains
Aleppo
R. Euphrates
Izmir
AEGEAN SEA
Rhodes
Iraklíon
Crete
Nicosia
CYPRUS
SYRIA
LEBANON
Beirut
Damascus
ISRAEL
Jerusalem
Amman
Dead Sea
JORDAN
Benghazi
Alexandria
EGYPT
El Giza
Cairo
SAUDI ARABIA

Bridges over the River Seine in Paris

Bridges over the
River Vltava in Prague

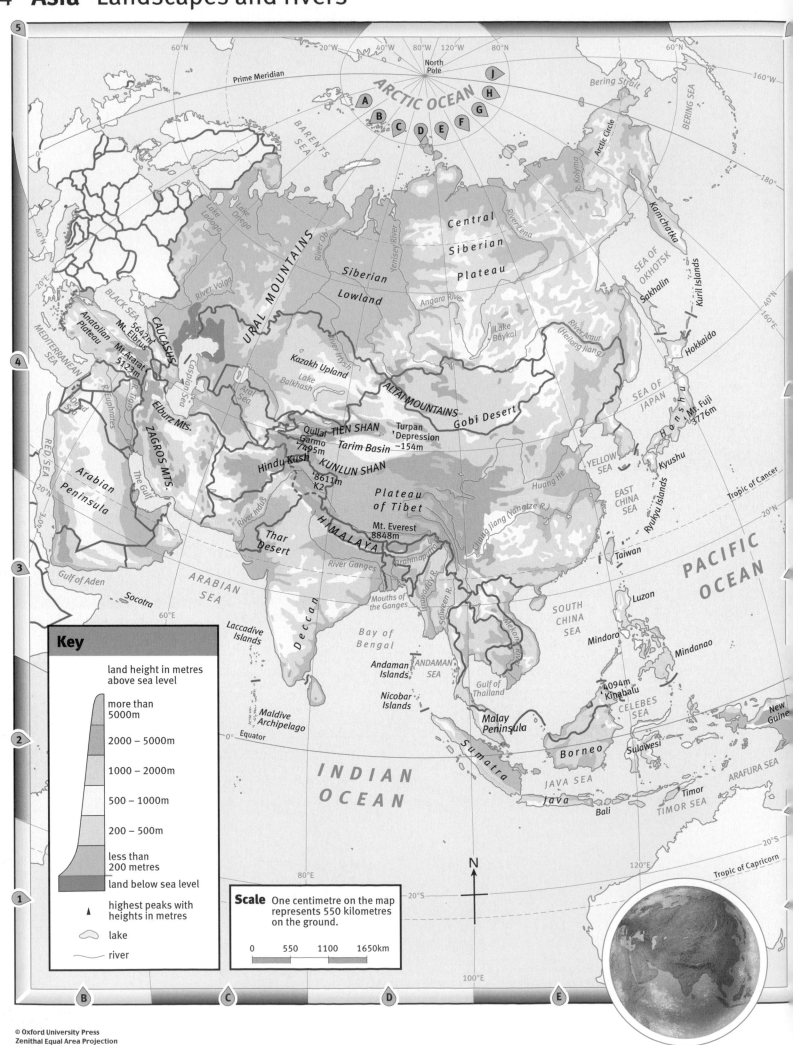

Key

land height in metres above sea level

more than 5000m

2000 – 5000m

1000 – 2000m

500 – 1000m

200 – 500m

less than 200 metres

land below sea level

▲ highest peaks with heights in metres

lake

river

Scale One centimetre on the map represents 550 kilometres on the ground.

0   550   1100   1650km

N

ARCTIC OCEAN
BARENTS SEA
BERING SEA
Bering Strait
SEA OF OKHOTSK
Kamchatka
Kuril Islands
Sakhalin
Hokkaido
SEA OF JAPAN
Honshu
Mt. Fuji 3776m
Kyushu
Ryukyu Islands
EAST CHINA SEA
Taiwan
YELLOW SEA
PACIFIC OCEAN
Luzon
Mindoro
Mindanao
SOUTH CHINA SEA
CELEBES SEA
4094m Kinabalu
Borneo
Sulawesi
JAVA SEA
Java
Bali
Timor
TIMOR SEA
New Guinea
ARAFURA SEA
Malay Peninsula
Gulf of Thailand
Mekong River
ANDAMAN SEA
Andaman Islands
Nicobar Islands
Bay of Bengal
Mouths of the Ganges
Irrawaddy R.
Salween R.
Brahmaputra
River Ganges
HIMALAYA
Mt. Everest 8848m
Plateau of Tibet
KUNLUN SHAN
8611m K2
Hindu Kush
Qullai Garmo 7495m
TIEN SHAN
Tarim Basin
Turpan Depression –154m
Gobi Desert
ALTAI MOUNTAINS
Kazakh Upland
Lake Balkhash
Aral Sea
Caspian Sea
Lake Baykal
Huang He
Chang Jiang (Yangtze R.)
River Amur (Heilong Jiang)
Angara River
Central Siberian Plateau
Siberian Lowland
River Ob
Yenisey River
River Lena
R. Kolyma
URAL MOUNTAINS
River Volga
Lake Ladoga
Lake Onega
BLACK SEA
CAUCASUS
Mt. Elbrus 5642m
Mt. Ararat 5123m
Anatolian Plateau
Elburz Mts.
ZAGROS MTS.
R. Euphrates
Tigris
Dead Sea
MEDITERRANEAN SEA
RED SEA
The Gulf
Arabian Peninsula
Gulf of Aden
Socotra
ARABIAN SEA
River Indus
Thar Desert
Deccan
Laccadive Islands
Maldive Archipelago
INDIAN OCEAN
Equator
Tropic of Cancer
Tropic of Capricorn
North Pole
Prime Meridian
Arctic Circle
Sumatra

A B C D E F G H J

© Oxford University Press
Zenithal Equal Area Projection

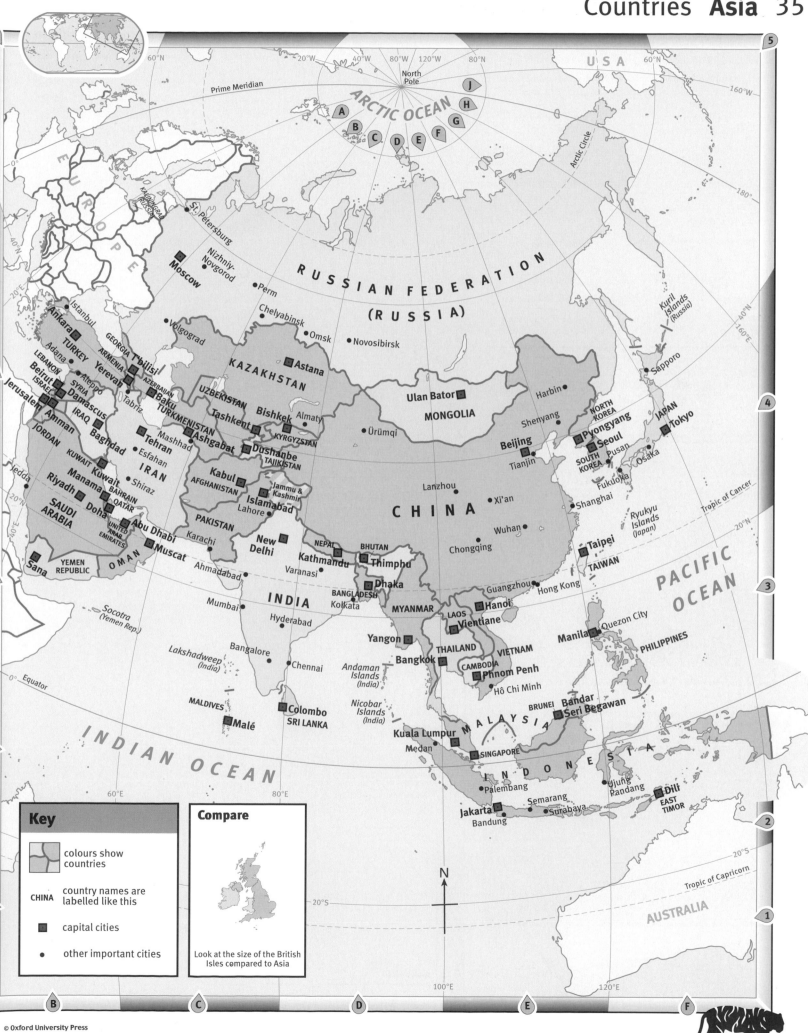

USA

ARCTIC OCEAN

North Pole

J
A
H
B
C G
D E F

60°N
20°W
40°W
80°W
120°W
80°N
60°N
160°W

Prime Meridian

Arctic Circle

180°

160°E

EUROPE

St. Petersburg

Nizhniy-Novgorod
Moscow
Perm
Chelyabinsk
Omsk
Novosibirsk

RUSSIAN FEDERATION
(RUSSIA)

Kuril Islands (Russia)

KALININGRAD (RUSSIA)

Istanbul
Ankara
TURKEY
Adana
GEORGIA T'bilisi
ARMENIA
Yerevan
AZERBAIJAN
Baku
TURKMENISTAN
Tabriz

Volgograd

Astana
KAZAKHSTAN

UZBEKISTAN
Tashkent
Bishkek
Almaty

Ulan Bator
MONGOLIA

Harbin

Sapporo

LEBANON
Beirut
SYRIA
Aleppo
ISRAEL
Jerusalem
Damascus
Amman
JORDAN
IRAQ
Baghdad
Tehran
Mashhad
Esfahan
IRAN
Shiraz

Ashgabat
Dushanbe
TAJIKISTAN

Ürümqi

Shenyang
NORTH KOREA
Pyongyang
Beijing
Tianjin
SOUTH KOREA
Seoul
Pusan
JAPAN
Tokyo
Osaka
Fukuoka

4

KUWAIT
Kuwait
Manama
BAHRAIN
Riyadh
Doha
QATAR
SAUDI
ARABIA
Abu Dhabi
UNITED ARAB EMIRATES
Muscat
OMAN

Kabul
AFGHANISTAN
Jammu & Kashmir
Islamabad
Lahore
PAKISTAN

Karachi

New Delhi

NEPAL
Kathmandu
Varanasi

Lanzhou
Xi'an
CHINA
Wuhan
Chongqing

Shanghai

Ryukyu Islands (Japan)

Tropic of Cancer
20°N

Jedda
YEMEN REPUBLIC
Sana

BHUTAN
Thimphu

Dhaka
BANGLADESH

Guangzhou
Hong Kong

Taipei
TAIWAN

PACIFIC OCEAN

3

Socotra (Yemen Rep.)

Ahmadabad

INDIA

Mumbai
Hyderabad

Kolkata

MYANMAR

LAOS
Hanoi
Vientiane

Manila
Quezon City
PHILIPPINES

Lakshadweep (India)

Bangalore
Chennai

Yangon

THAILAND
Bangkok
CAMBODIA
VIETNAM
Phnom Penh
Hô Chi Minh

Equator

Andaman Islands (India)

MALDIVES
Male
Colombo
SRI LANKA

Nicobar Islands (India)

BRUNEI
Bandar Seri Begawan

INDIAN OCEAN

Kuala Lumpur
Medan
MALAYSIA
SINGAPORE

INDONESIA

Ujung Pandang
Dili
EAST TIMOR

60°E
80°E

Palembang
Jakarta
Bandung
Semarang
Surabaya

2

N

20°S

Tropic of Capricorn

AUSTRALIA

1

100°E
120°E

Key

colours show countries

CHINA country names are labelled like this

▣ capital cities

• other important cities

Compare

Look at the size of the British Isles compared to Asia

B
C
D
E
F

Key

——	country boundary
- - -	disputed boundary
——	motorway or main road
——	railway
⊕	main airport
∿	river
⌒	lake

towns and cities

■	capital cities
o	largest towns
•	other large towns

land height
above sea level in metres

more than 5000m

2000 – 5000m

1000 – 2000m

500 – 1000m

200 – 500m

less than 200 metres

land below sea level

▲ highest peaks with heights in metres

Scale One centimetre on the map represents 125 kilometres on the ground.

0   125   250   375km

Traffic jam in Rajasthan

Rush hour in Jaipur

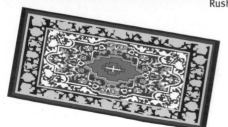

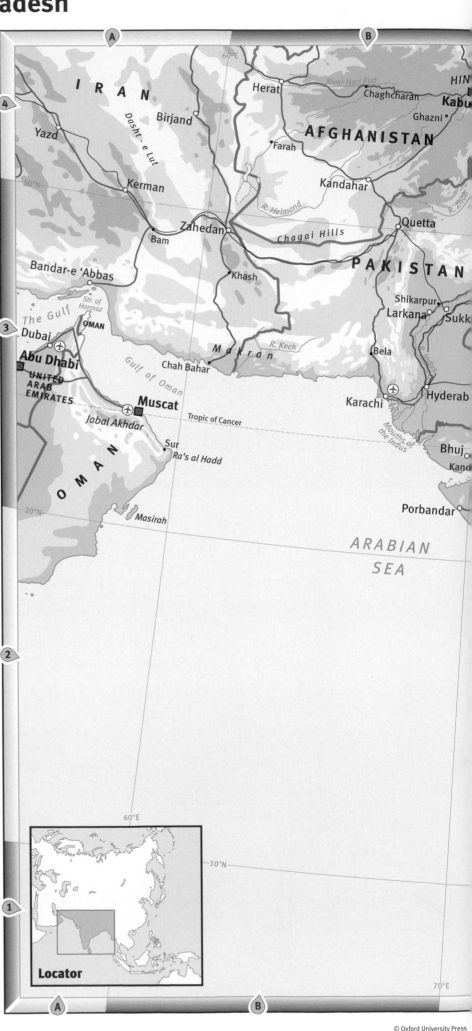

© Oxford University Press
Conical Orthomorphic Projection

PAMIRS
TAJIKISTAN
Khorog
7690m
Gilgit

HINDU KUSH
Peshawar
Srinagar
JAMMU
AND
KASHMIR
Leh
Islamabad
Rawalpindi
R. Indus
Jammu
R. Jhelum
Gujranwala
Lahore
Amritsar
Faisalabad
Chenab
Multan
Dera Ghazi Khan
River Sutlej
Ludhiana
Chandigarh
Bahawalpur
Rahimyar Khan

K2 (Qogir Feng, Godwin Austen) 8611m
R. Indus
Rutog

C H I N A

Jinsha Jiang (Yangtze R.)
Lancang Jiang (Mekong R.)

Nu Jiang (Salween R.)

Lhasa
Nyingchi

Yarlung Zangbo (Tsangpo R.)

H I M A L A Y A

Dibrugarh

Dehra Dun
Meerut
R. Yamuna
New Delhi
Delhi
Bareilly
Bikaner
Jaipur
Agra
River Ganges
R. Ghaghara
Lucknow
Gorakhpur
N E P A L
Annapurna 8091m
Mount Everest 8848m
Kathmandu
Darjiling
Thimphu
BHUTAN
Guwahati
Nagaon
Shillong
Brahmaputra R.
Imphal
River Chindwin

Thar Desert

Jodhpur
R. Banas
Gwalior
Kanpur
R. Chambal
Kota
Jhansi
Allahabad
Varanasi
R. Gomti
Patna
Bhagalpur
R. Ganges
R. Son
Muzaffarpur

Gandhi Sagar
I N D I A
Murwara
Dhanbad
Asanol
BANGLADESH
Dhaka
Tropic of Cancer

Ahmadabad
Bhopal
Jabalpur
Jamshedpur
Kolkata
Khulna
Chittagong
Monywa
Vadodara
Indore
R. Narmada
Kharagpur
Mouths of the Ganges
Mandalay
Rajkot
R. Tapi
Bilaspur
Hirakud Reservoir
Sambalpur
MYANMAR (BURMA)
Bhavnagar
Bharuch
Burhanpur
Raipur
Surat
Dhule
Amravati
Nagpur
Sittwe
Arakan Yoma
Nashik
R. Mahanadi
Cuttack
21°N
Gulf of Khambhat
Aurangabad
Chandrapur
R. Godavari
Sandoway
Pye
Irrawaddy R.

Mumbai
Pune
Nizamabad
R. Indravati
Brahmapur
Bay of Bengal
Yangon
Deccan
R. Godavari
Bassein
Solapur
R. Bhima
Vishakhapatnam
Kolhapur
WESTERN GHATS
Hyderabad
Bijapur
Raichur
R. Krishna
Rajahmundry
Mouths of the Irrawaddy
Belgaum
Vijayawada
EASTERN GHATS
Bellary
R. Penner
Nellore
Andaman Islands

Mangalore
Bangalore
Vellore
Chennai
Mysore
Port Blair
Pondicherry
ANDAMAN SEA
Laccadive Islands
Calicut
Salem
Coimbatore
Tiruchchirappalli
I N D I A N   O C E A N
10°N
Cochin
Madurai
Jaffna
Quilon
Trivandrum
SRI LANKA
Trincomalee
Nagercoil
Batticaloa
Puttalam
Colombo
Kandy
Badulla
Galle

80°E
90°E

RUSSIAN FEDERATION (RUSSIA)

Grid references (top): A, B, C, D, E
Longitude: 80°E, 90°E, 100°E, 110°E, 120°E

Pavlodar
Barnaul
Biysk
Astana
River Ob
Rubtsovsk
Semipalatinsk
Karaganda
Ust'-Kamenogorsk
Zyryanovsk
KAZAKHSTAN
Ayaguz
Taldykorgan
Lake Zaysan
Lake Alakol
Lake Balkhash
Altay
Ulaangom
Hovd
ALTAI MOUNTAINS
Yenisey River
Mur
Hövsgöl Nuur
Selenge River
Angarsk
Irkutsk
Lake Baykal
Ulan-Ude
Chita
Borzya
Argun R. (Ergun H
Manzhouli
Choybalsan
Ulan Bator
M O N G O L I A
G o b i   D e s e r t
Saynshand
Erenhot
Almaty
Yining
Bishkek
KYRGYZSTAN
Issyk
TIEN SHAN
Ürümqi
Turpan
Turpan Depression −154m
Hami
Kashi
Tarim He
Tarim Pendi
Hotan He
Lop Nur
Anxi
Yumen
Qilian Shan
Hohhot
Jining
Zhangjiako
Baotou
Wuhai
Datong
Beiji
Tangshan
Tianjin
Yinchuan
Great Wall
Huang He
Shijiazhua
K2 (Qogir Feng) 8611m
Altun Shan
Kunlun Shan
JAMMU AND KASHMIR
R. Indus
Rutog
C H I N A
Golmud
Qinghai Hu
Xining
Lanzhou
Taiyuan
Handan
Dezhou
Jinan
H
Plateau of Tibet
Huang He
Baoji
Changzhi
Zhengzhou
Luoyang
Xuzhou
Suzhou
Bengbu
Hefei
Wen He
Xi'an
Dehra Dun
M
Batang
Chengdu
Chang Jiang (Yangtze River)
Xiangfan
Wuhan
Jingdez
Bareilly
NEPAL
Annapurna 8091m
Lhaze
Lhasa
Yarlung Zangbo (Tsangpo)
Chongqing
Changde
Dongting Hu
Poyang Hu
Nanchang
Lucknow
Darjiling
Mt. Everest 8848m
A Y A
Kathmandu
Thimphu
BHUTAN
Dibrugarh
Neijiang
Yibin
Zunyi
Changsha
Zhuzhou
Ji'an
Kanpur
Gorakhpur
Muzaffarpur
Shiliguri
Brahmaputra R.
Guiyang
Shaoyang
Hengyang
Allahabad
Varanasi
Patna
Bhagalpur
Shillong
Dali
Kunming
Duyun
Guilin
Shaoguan
Ganzh
Ganges
Murwara
Tropic of Cancer
Dhanbad
BANGLADESH
Imphal
R. Chindwin
Nan Ling
Meizh
Jabalpur
Jamshedpur
Dhaka
Liuzhou
Wuzhou
Guangzho
INDIA
Bilaspur
Kharagpur
Khulna
Chittagong
Monywa
Nanning
Macao
Hong Kong
Raipur
Kolkata
Mouths of the Ganges
MYANMAR
(BURMA)
Mandalay
Salween R.
Mekong R.
Song Koi
Lao Cai
Pingxiang
Zhanjiang
Haikou
Cuttack
Sittwe
Arakan Yoma
Irrawaddy R.
Kengtung
Phongsali
Hanoi
Hai Phong
Hainan Dao
Brahmapur
Bay of Bengal
Pye
Chiang Mai
Louangphrabang
Thanh Hoa
Vinh
Sanya
Vishakhapatnam
Bassein
Pegu
Udon Thani
Vientiane
LAOS
VIETNAM
Hue
Da Nang
SOUTH CHINA SEA
Mouths of the Irrawaddy
Yangon
Moulmein
THAILAND

Latitude markers: 50°N, 40°N, 30°N, 20°N
Longitude (bottom): 90°E, 100°E, 110°E

Mt. Fuji is Japan's highest peak

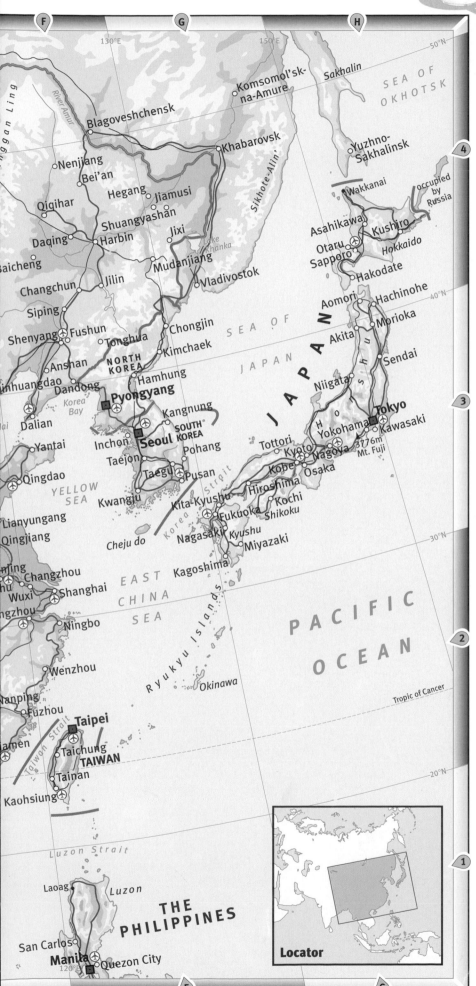

SEA OF OKHOTSK

River Amur
Komsomol'sk-na-Amure
Sakhalin
Blagoveshchensk
Nenjiang
Bei'an
Hegang
Jiamusi
Khabarovsk
Qiqihar
Shuangyashan
Daqing
Harbin
Jixi
Yuzhno-Sakhalinsk
Baicheng
Mudanjiang
Sikhote-Alin
occupied by Russia
Changchun
Jilin
Vladivostok
Wakkanai
Siping
Asahikawa
Kushiro
Shenyang
Fushun
Tonghua
Chongjin
Otaru
Sapporo
Hokkaido
Anshan
Kimchaek
Hakodate
NORTH KOREA
Hamhung
Aomori
Hachinohe
Dandong
Pyongyang
Kangnung
SEA OF JAPAN
Akita
Morioka
Korea Bay
Dalian
Seoul
SOUTH KOREA
Pohang
Niigata
Sendai
Yantai
Inchon
Taejon
Qingdao
Taegu
Pusan
Tottori
Kyoto
Yokohama
Tokyo
Kawasaki
YELLOW SEA
Kwangju
Kobe
Nagoya
3776m
Mt. Fuji
Kita-Kyushu
Hiroshima
Kochi
Osaka
Cheju do
Nagasaki
Kyushu
Shikoku
Fukuoka
Miyazaki
EAST CHINA SEA
Kagoshima
Changzhou
Shanghai
Wuxi
Ningbo
Ryukyu Islands
PACIFIC OCEAN
Wenzhou
Okinawa
Tropic of Cancer
Nanping
Fuzhou
Taipei
Taichung
TAIWAN
Tainan
Kaohsiung
Luzon Strait
Laoag
Luzon
THE PHILIPPINES
San Carlos
Manila
Quezon City

Shopping in Shanghai, China

Key

		land height
——	country boundary	above sea level in metres
---	disputed boundary	
——	motorway or main road	more than 5000m
—	railway	2000 – 5000m
⊕	main airport	1000 – 2000m
～	river	500 – 1000m
⌒	lake	200 – 500m

towns and cities

■	capital cities	less than 200 metres
○	largest towns	land below sea level
•	other large towns	▲ highest peaks with heights in metres

Scale

One centimetre on the map represents 180 kilometres on the ground.

0	180	360	540km

Locator

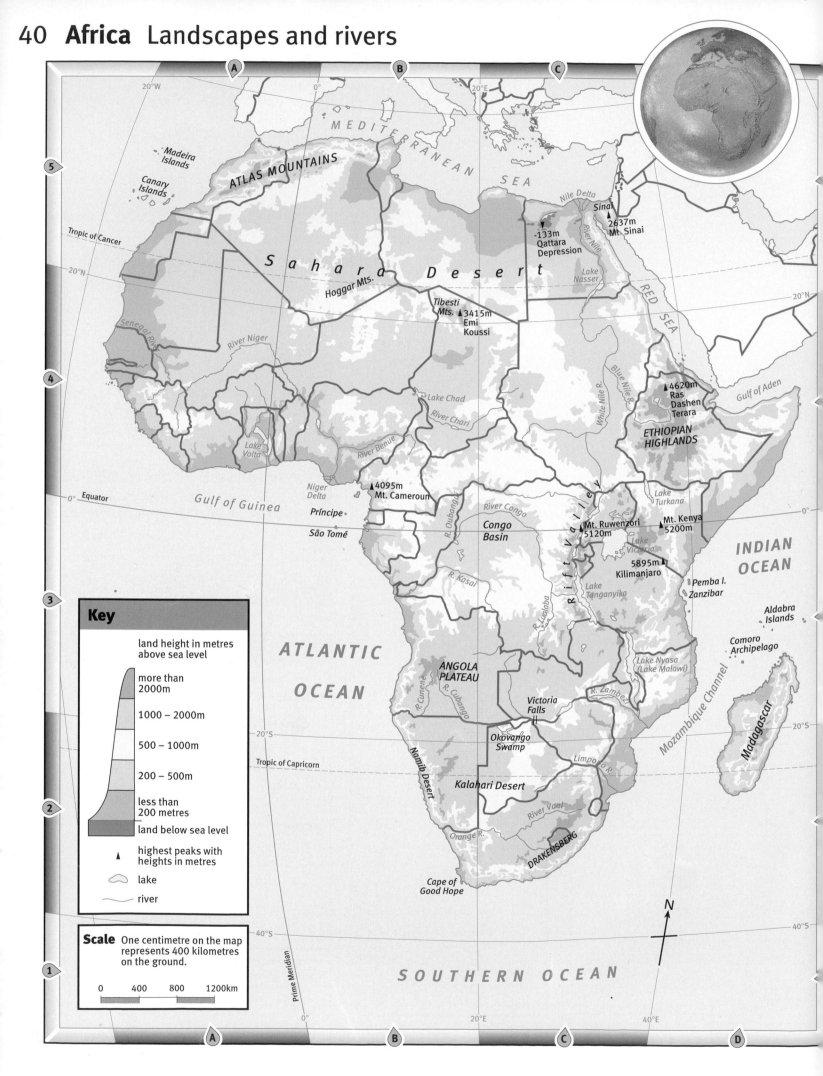

Key

land height in metres
above sea level

more than
2000m

1000 – 2000m

500 – 1000m

200 – 500m

less than
200 metres

land below sea level

▲ highest peaks with
heights in metres

lake

river

Scale One centimetre on the map
represents 400 kilometres
on the ground.

0   400   800   1200km

MEDITERRANEAN SEA

ATLAS MOUNTAINS

Madeira
Islands

Canary
Islands

Tropic of Cancer

20°N

S a h a r a   D e s e r t

Hoggar Mts.

Senegal River

River Niger

Tibesti
Mts. ▲3415m
Emi
Koussi

Nile Delta

Sinai
▲2637m
Mt. Sinai

-133m
Qattara
Depression

River Nile

Lake
Nasser

RED SEA

Lake Chad

River Chari

Lake
Volta

River Benue

▲4095m
Mt. Cameroun

Niger
Delta

Gulf of Guinea

Príncipe

São Tomé

0° Equator

White Nile R.

Blue Nile R.

▲4620m
Ras
Dashen
Terara

ETHIOPIAN
HIGHLANDS

Gulf of Aden

Lake
Turkana

R. Oubangui

River Congo

Congo
Basin

R. Kasai

R. Lualaba

Rift Valley

Mt. Ruwenzori
5120m

Lake
Victoria

▲Mt. Kenya
5200m

5895m
Kilimanjaro

Lake
Tanganyika

Pemba I.
Zanzibar

INDIAN
OCEAN

ATLANTIC

OCEAN

ANGOLA
PLATEAU

R. Cunene

R. Cubango

Lake Nyasa
(Lake Malawi)

Aldabra
Islands

Comoro
Archipelago

R. Zambezi

Victoria
Falls

Okovango
Swamp

Namib Desert

20°S

Tropic of Capricorn

Kalahari Desert

Limpopo R.

Mozambique Channel

Madagascar

River Vaal

Orange R.

DRAKENSBERG

Cape of
Good Hope

N

Prime Meridian

40°S

SOUTHERN OCEAN

Zenithal Equal Area Projection   © Oxford University P

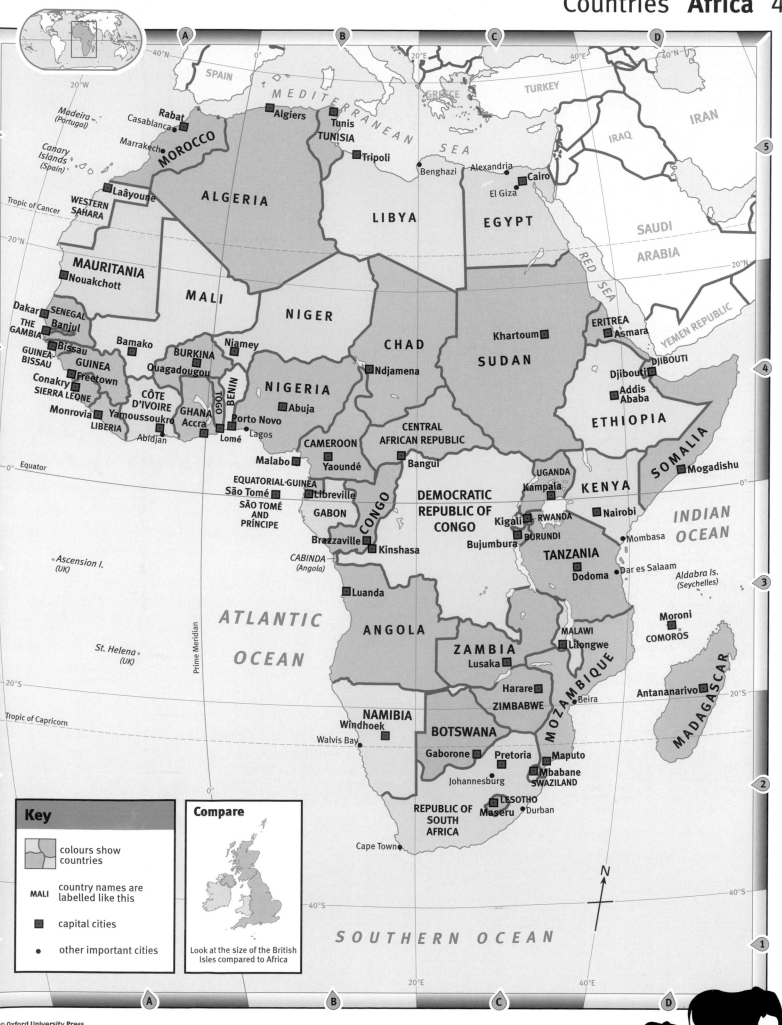

SPAIN

M E D I T E R R A N E A N   S E A

GREECE   TURKEY

IRAQ   IRAN

Madeira
(Portugal)

Rabat
Casablanca
Marrakech

MOROCCO

Algiers

Tunis
TUNISIA

Tripoli

Benghazi   Alexandria
Cairo
El Giza

Canary
Islands
(Spain)

Laâyoune

Tropic of Cancer

**WESTERN
SAHARA**

ALGERIA

LIBYA

EGYPT

SAUDI
ARABIA

20°N

MAURITANIA
Nouakchott

MALI

NIGER

Khartoum

SUDAN

RED SEA

ERITREA
Asmara

YEMEN REPUBLIC

Dakar **SENEGAL**
THE   Banjul
GAMBIA
Bissau
GUINEA-
BISSAU
Conakry
GUINEA
Freetown
SIERRA LEONE
Monrovia Yamoussoukro
LIBERIA
Abidjan

Bamako

Niamey

CHAD

Ndjamena

DJIBOUTI
Djibouti

Addis
Ababa

BURKINA
Ouagadougou

**CÔTE
D'IVOIRE**
GHANA
Accra
Lomé
TOGO
BENIN

NIGERIA

Abuja

Porto Novo
Lagos

**CENTRAL
AFRICAN REPUBLIC**

ETHIOPIA

CAMEROON
Malabo
Yaoundé
EQUATORIAL GUINEA
São Tomé
**SÃO TOMÉ
AND
PRÍNCIPE**

Bangui

Libreville
GABON

**DEMOCRATIC
REPUBLIC OF
CONGO**

UGANDA
Kampala

KENYA

SOMALIA
Mogadishu

Nairobi

0° Equator

CONGO
Brazzaville
Kinshasa

Kigali **RWANDA**
Bujumbura **BURUNDI**

Mombasa

**INDIAN
OCEAN**

CABINDA
(Angola)

TANZANIA
Dodoma
Dar es Salaam

Aldabra Is.
(Seychelles)

Ascension I.
(UK)

Luanda

**ATLANTIC

OCEAN**

St. Helena
(UK)

Prime Meridian

ANGOLA

ZAMBIA
Lusaka

MALAWI
Lilongwe

Moroni
COMOROS

MADAGASCAR

20°S

Harare
ZIMBABWE

Beira

MOZAMBIQUE

Antananarivo

Tropic of Capricorn

NAMIBIA
Windhoek
Walvis Bay

BOTSWANA

Gaborone Pretoria
Johannesburg

Maputo
Mbabane
SWAZILAND

LESOTHO
Maseru Durban

**REPUBLIC OF
SOUTH
AFRICA**

Cape Town

N

Key

colours show
countries

MALI country names are
labelled like this

capital cities

other important cities

Compare

Look at the size of the British
Isles compared to Africa

S O U T H E R N   O C E A N

20°E   40°E

5

4

3

2

1

© Oxford University Press

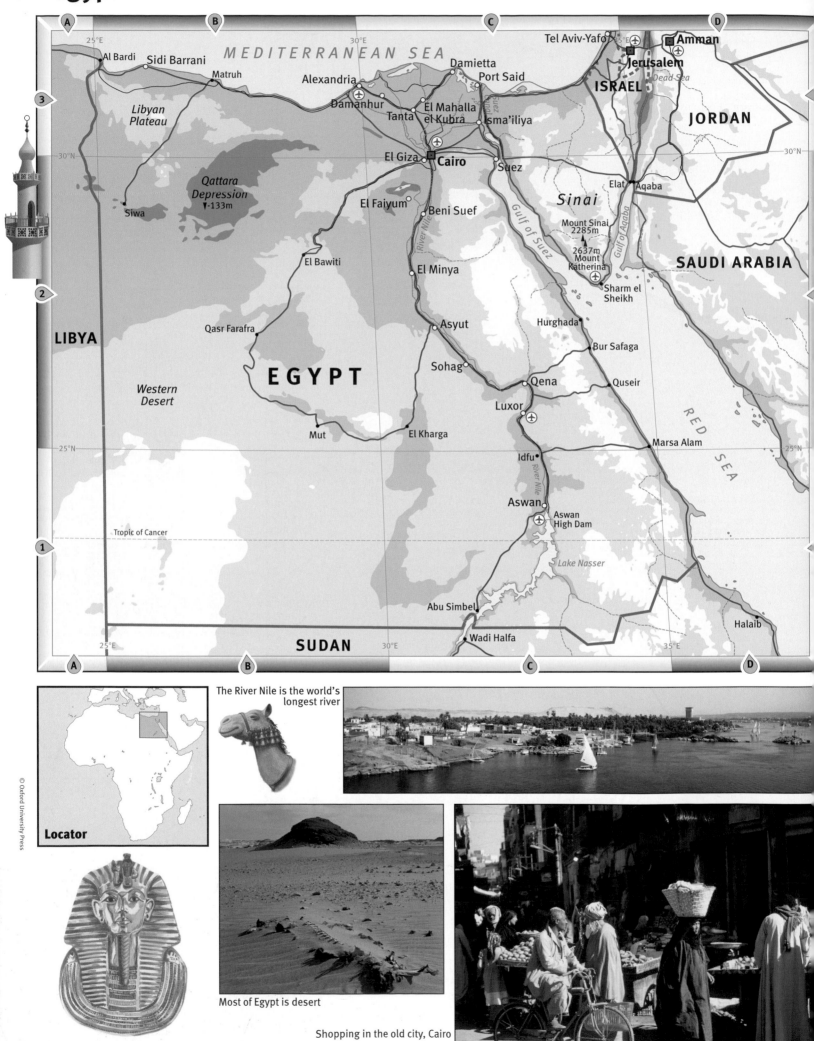

MEDITERRANEAN SEA

Al Bardi · Sidi Barrani · Matruh · Alexandria · Damietta · Port Said · Tel Aviv-Yafo · Amman · Jerusalem · ISRAEL · JORDAN · Dead Sea

Libyan Plateau

Damanhur · El Mahalla el Kubra · Tanta · Isma'iliya

El Giza · Cairo · Suez · Sinai · Elat · Aqaba

Qattara Depression ▼-133m

Siwa

El Faiyum · Beni Suef

Mount Sinai 2285m · 2637m Mount Katherina

SAUDI ARABIA

El Bawiti · El Minya · Sharm el Sheikh

LIBYA

Qasr Farafra · Asyut · Hurghada

EGYPT · Bur Safaga

Western Desert · Sohag · Qena · Quseir

Luxor

Mut · El Kharga · Marsa Alam

Idfu

Tropic of Cancer

Aswan · Aswan High Dam

Lake Nasser

Abu Simbel

RED SEA

Halaib

SUDAN · Wadi Halfa

Locator

The River Nile is the world's longest river

Most of Egypt is desert

Shopping in the old city, Cairo

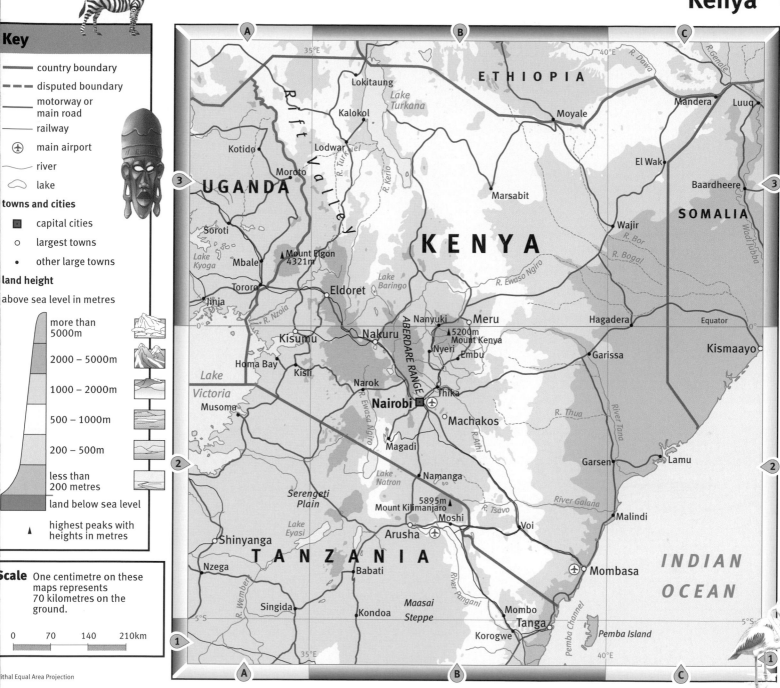

Key

- —— country boundary
- – – – disputed boundary
- —— motorway or main road
- —— railway
- ⊕ main airport
- ◠ river
- ◠ lake

towns and cities

- ▣ capital cities
- ○ largest towns
- • other large towns

land height

above sea level in metres

- more than 5000m
- 2000 – 5000m
- 1000 – 2000m
- 500 – 1000m
- 200 – 500m
- less than 200 metres
- land below sea level
- ▲ highest peaks with heights in metres

Scale One centimetre on these maps represents 70 kilometres on the ground.

| 0 | 70 | 140 | 210km |

ithal Equal Area Projection

Map labels

ETHIOPIA
UGANDA
KENYA
SOMALIA
TANZANIA
INDIAN OCEAN

Rift Valley
Lake Turkana
Lokitaung
Kalokol
Kotido
Lodwar
Moroto
Soroti
Mbale
Mount Elgon 4321m
Tororo
Jinja
Lake Kyoga
Eldoret
R. Nzoia
Kisumu
Nakuru
Homa Bay
Kisii
Lake Victoria
Musoma
Narok
R. Ewaso Ngiro
Nairobi ⊕
Magadi
Namanga
Lake Natron
Serengeti Plain
Lake Eyasi
Mount Kilimanjaro 5895m ▲
Moshi
Arusha ⊕
Shinyanga
Nzega
Babati
Singida
Kondoa
Maasai Steppe
Korogwe
Tanga
Mombo
Pemba Channel
Pemba Island

Lake Baringo
Nanyuki
Meru
ABERDARE RANGE
Mount Kenya 5200m ▲
Nyeri
Embu
Thika
Machakos
R. Athi
R. Tsavo
River Pangani
River Galana
R. Thua
River Tana
Garissa
Hagadera
Equator
Kismaayo
Garsen
Lamu
Malindi
Voi
Mombasa ⊕

Moyale
Marsabit
R. Ewaso Ngiro
R. Dawa
R. Genale
Mandera
Luuq
El Wak
Baardheere
Wajir
R. Bor
R. Bogal
Wadi Jubba

35°E 40°E
0° 5°S

The Maasai people herd cattle in central Kenya

Kilimanjaro is Africa's highest mountain

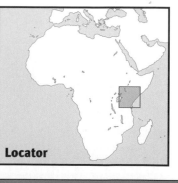

From Mombasa to Nairobi by road takes about 8 hours

Locator

© Oxford University Press

Key

land height in metres above sea level

more than 2000m

1000 – 2000m

500 – 1000m

200 – 500m

less than 200 metres

land below sea level

▲ highest peaks with heights in metres

lake

river

Scale One centimetre on the map represents 400 kilometres on the ground.

0   400   800   1200km

Oblique Mercator Projection   © Oxford University Press

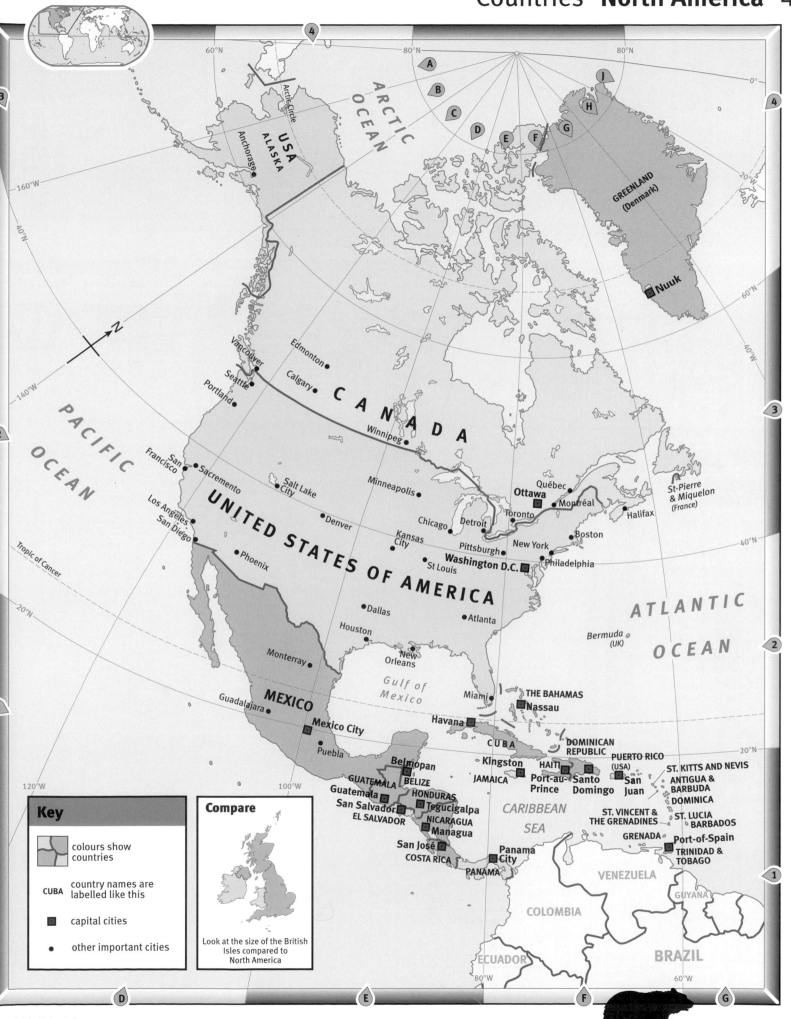

ARCTIC OCEAN

60°N 80°N 80°N

A
B
C
D E F G H J
0°

ARCTIC OCEAN

Arctic Circle

USA
ALASKA
Anchorage

160°W

40°N

GREENLAND
(Denmark)

20°W

Nuuk

60°N

Vancouver
Edmonton

40°W

Seattle
Calgary
Portland

C A N A D A

140°W

Winnipeg

PACIFIC
OCEAN

San
Francisco
Sacramento
Salt Lake
City
Minneapolis

Québec
Ottawa
Montréal

St-Pierre
& Miquelon
(France)

Los Angeles
San Diego

UNITED STATES OF AMERICA

Chicago Detroit
Toronto
Halifax

Denver
Kansas
City
Pittsburgh
New York
Boston

40°N

Phoenix
St Louis
Washington D.C.
Philadelphia

ATLANTIC

Tropic of Cancer

Dallas
Atlanta

20°N

Houston

OCEAN

Monterray

New
Orleans
Bermuda
(UK)

2

Guadalajara
MEXICO

Gulf of
Mexico

Miami

THE BAHAMAS
Nassau

Mexico City

Havana

CUBA

DOMINICAN
REPUBLIC

20°N

Puebla

Belmopan

Kingston

PUERTO RICO
(USA)

HAITI

ST. KITTS AND NEVIS

GUATEMALA
BELIZE
JAMAICA
Port-au-
Prince
Santo
Domingo
San
Juan

ANTIGUA &
BARBUDA
DOMINICA

Guatemala
HONDURAS

San Salvador
Tegucigalpa
EL SALVADOR
NICARAGUA
Managua

CARIBBEAN
SEA

ST. VINCENT &
THE GRENADINES
ST. LUCIA
BARBADOS

GRENADA
Port-of-Spain

San José
COSTA RICA
Panama
City
PANAMA

TRINIDAD &
TOBAGO

1

VENEZUELA

GUYANA

COLOMBIA

ECUADOR
80°W
BRAZIL
60°W

120°W 100°W

Key

colours show
countries

CUBA country names are
labelled like this

capital cities

other important cities

Compare

Look at the size of the British
Isles compared to
North America

The Rocky Mountains stretch over 1000 miles

Evening traffic, Los Angeles

Locator

Key

——	country boundary
- - -	disputed boundary
——	motorway or main road
—	railway
✈	main airport
~~	river
⬭	lake

towns and cities

▣	capital cities
○	largest towns
•	other large towns

land height

above sea level in metres

more than 5000m	
2000 – 5000m	
1000 – 2000m	
500 – 1000m	
200 – 500m	
less than 200 metres	
land below sea level	
▲	highest peaks with heights in metres

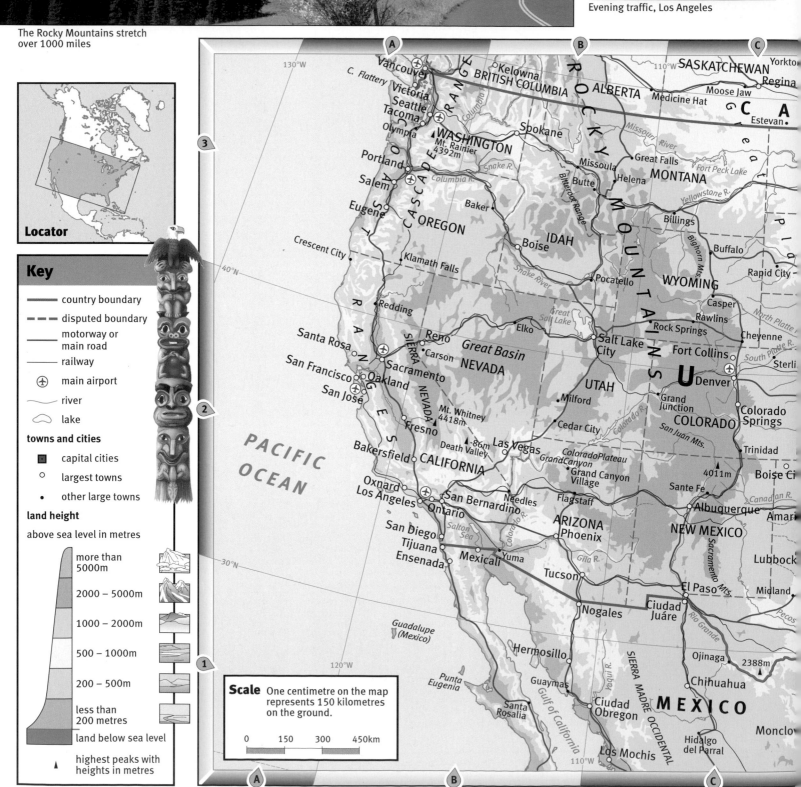

Scale One centimetre on the map represents 150 kilometres on the ground.

0   150   300   450km

Wyoming has fewer people than any other state in the USA

Manhattan skyline, New York City

MANITOBA
Winnipeg
Manitoba
Lake Winnipeg
ONTARIO
Longlac
Minot
Upper Red L.
Grand Forks
Lower Red L.
Bemidji
NORTH DAKOTA
Bismarck
Fargo
MINNESOTA
Duluth
Ironwood
WISCONSIN
St. Paul
Minneapolis
SOUTH DAKOTA
Pierre
Mitchell
Albert Lea
Madison
Sioux Falls
Sioux City
IOWA
Cedar Rapids
Milwaukee
Des Moines
Iowa City
NEBRASKA
North Platte
Platte R.
Omaha
Lincoln
Springfield
ILLINOIS
Bloomington
Missouri River
S
KANSAS
Salina
Topeka
Dodge City
Wichita
Arkansas R.
Kansas City
Jefferson City
MISSOURI
Plateau
Ozark
Springfield
Tulsa
White R.
Oklahoma
Fort Smith
OKLAHOMA
Little Rock
Oklahoma City
ARKANSAS
Wichita Falls
Red River
Texarkana
MISSISSIPPI
Meridian
Fort Worth
Dallas
Shreveport
Jackson
Abilene
LOUISIANA
Colorado R.
TEXAS
Huntsville
Lafayette
Baton Rouge
Edwards Plateau
Austin
Houston
New Orleans
Del Rio
San Antonio
Galveston
Mississippi Delta
Piedras Negras
Nuevo Laredo
Laredo
Corpus Christi
Rio Grande
Gulf of Mexico
Reynosa
Monterrey
Matamoros

Lake Superior
Thunder Bay
The Great Lakes
Michipicoten
Marquette
Sault Ste. Marie
Sudbury
North Bay
Ottawa
Lake Huron
Traverse City
MICHIGAN
Green Bay
Grand Rapids
Lake Michigan
London
Toronto
Lake Ontario
Rochester
Lansing
Detroit
Lake Erie
Cleveland
Toledo
Akron
Fort Wayne
OHIO
INDIANA
Columbus
A
Indianapolis
Cincinnati
Chicago
Springfield
Louisville
Ohio R.
Frankfort
Lexington
St. Louis
KENTUCKY
Bowling Green
Nashville
Knoxville
TENNESSEE
Chattanooga
Memphis
Tennessee R.
Greenville
Birmingham
ALABAMA
Columbus
Montgomery
GEORGIA
Macon
Alabama R.
Mobile
Tallahassee
Daytona Beach
Orlando
C. Canaveral
Tampa
FLORIDA
St. Petersburg
L. Okeechobee
West Palm Beach
Naples
Miami
Florida Keys
Straits of Florida

Réservoir Gouin
Rivière-du-Loup
Québec
Val-d'Or
QUÉBEC
St. Lawrence
Montréal
Sherbrooke
Kingston
NEW YORK
VERMONT
NEW HAMPSHIRE
Concord
Syracuse
Albany
MASS.
Buffalo
Hartford
CONN.
St. Catharines
Scranton
PENNSYLVANIA
New York
Newark
Trenton
Pittsburgh
Philadelphia
Harrisburg
NEW JERSEY
Dover
DELAWARE
Baltimore
Washington D.C.
MARYLAND
WEST VIRGINIA
Charleston
Richmond
APPALACHIAN MTS.
VIRGINIA
Norfolk
Greensboro
Raleigh
Cape Hatter
NORTH CAROLINA
Charlotte
Florence
Wilmington
SOUTH CAROLINA
Columbia
Charleston
Savannah River
Savannah
Jacksonville

Presque Isle
MAINE
Bangor
Augusta
Portland
Boston
Cape Cod
Providence
R.I.
Chesapeake Bay

NEW BRUNSWICK
Saint John
Bay of Fundy
NOVA SCOTIA
Halifax
Yarmouth

ATLANTIC OCEAN

Daytona Beach
Freeport
Grand Bahama
Great Abaco
THE BAHAMAS
Nassau
New Providence I.
Eleuthera
Cat I.
Andros
Tropic of Cancer
Long Island

100°W
90°W
80°W
40°N
30°N

© Oxford University Press

Key

——	country boundary
– – –	disputed boundary
——	motorway or main road
——	railway
✈	main airport
∿	river
◠	lake

land height

above sea level in metres

	more than 5000m
	2000 – 5000m
	1000 – 2000m
	500 – 1000m
	200 – 500m
	less than 200 metres
	land below sea level
▲	highest peaks with heights in metres

towns and cities

■	capital cities
○	largest towns
•	other large towns

Locator

Fishing boats in St. Lucia

Catamarans in the Virgin Islands

FLORIDA

Daytona Beach · 80°W · 75°W
Orlando ✈
Cape Canaveral
Tampa
St. Petersburg
L. Okeechobee
West Palm Beach
Freeport · Grand Bahama · Marsh Harbour · Great Abaco
Miami ✈
New Providence Island · Governor's Harbour · Eleuthera
25°N
Key West · Florida Keys · Andros Town · **Nassau** · Cat Island · San Salvador
Straits of Florida · Andros · **THE BAHAMAS**
Tropic of Cancer · Great Exuma · Long Island · Crooked I.
Archipiélago de Sabana
Havana ✈ · Matanzas · Acklins Island
Pinar del Río · Güines · Sagua la Grande
Le Fé · Cienfuegos · Santa Clara · Archipiélago de Camagüey
Cabo San Antonio · Sancti Spíritus · Morón · **CUBA** · Nuevitas
Isla de la Juventud · Trinidad · Ciego de Ávila · Camagüey · Victoria de las Tunas
Holguín
G r e a t e r · Bayamo · Guantánamo
Manzanillo · Sierra Maestra · Santiago de Cuba
20°N · Windwa
Cayman Islands (UK)
George Town ■ · Grand Cayman · Jérémie
Montego Bay ✈
South Negril Point · **JAMAICA** ✈
Black River · Spanish Town · **Kingston** ■

C A R I B B E

15°N

80°W

75°W

Scale
One centimetre on the map represents 80 kilometres on the ground.

0	80	160	240km

Main map

ATLANTIC

OCEAN

70°W

65°W

25°N

Tropic of Cancer

20°N

Mayaguana

Caicos Passage

Caicos Islands

Little
nagua I.

Turks and
Caicos Is.
(UK)

■ **Grand Turk**

Turks I.
Passage

Turks
Islands

Great
Inagua

H i s p a n i o l a

W e s t

Port-de-Paix

Cap Haïtien

Santiago

San Francisco

DOMINICAN
REPUBLIC

San
Juan

Charlotte
Amalie

Road
Town

Virgin Is.
(UK)

The Valley

Anguilla (UK)

Saint Martin (Fr.)

ANTIGUA AND
BARBUDA

Barbuda

I n d i e s

e de la
onâve

HAITI

Cordillera Central

La Vega

San Pedro

Aguadilla

Mayagüez

Caguas

Virgin Is.
(USA)

St. Maarten
(Neths)

Codrington

St. Kitts

Antigua

Nevis

St. John's

I s l a n d s

60°W

Port-au-
Prince

3175m

Santo
Domingo

La
Romana

Ponce

Puerto Rico
(USA)

St. Croix
(USA)

Basseterre
ST. KITTS
AND NEVIS

Plymouth

Montserrat
(UK)

Grande Terre

Guadeloupe (Fr.)

Pointe-á-Pitre

s Cayes

Jacmel

Barahona

Mona Passage

A n t i l l e s

Cabo Beata

L e s s e r

Marie Galente

Basse-Terre

DOMINICA

15°N

Roseau

Martinique (Fr.)

Fort-de-France

Castries
ST. LUCIA

Vieux Fort

BARBADOS
Bridgetown

S E A

L e s s e r A n t i l l e s

Aruba
(Neths.)

Curacao
(Neths.)

Bonaire
(Neths.)

St. Vincent

ST. VINCENT AND
THE GRENADINES

Kingstown

Oranjestad

Willemstad

Netherlands Antilles

Punta Gallinas

Punto Fijo

Golfo de
Venezuela

70°W

65°W

Isla Margarita

Carúpano

St. George's

GRENADA

Tobago

Port-of-
Spain

TRINIDAD
AND
TOBAGO

Trinidad

San Fernando

W i n d w a r d   I s l a n d s

Inset map: St. Lucia

St. Lucia

61°W

Saint Lucia Channel

Pte. du Cap

Pigeon Pt.

Gros Islet

Choc Bay

Monchy

Cape
Marquis

Grande Rivière

■ **Castries**

• **Marquis**

Babonneau

Grande
Anse
Bay

14°N

La Croix
Mairgot

La Sorcière
675m

Marigot

R. Roseau

Grande
Rivière

La Caye

Anse la Raye

Canaries

Dennery

R. Mabouya

Mt. Gimie
950m

Praslin Bay

Soufrière

Mon Repos

R. Troumassé

Micoud

Gros Piton
798m

R. Canelles

Desruisseaux

Choiseul

Augier

Saltibus Pt.

Laborie

Vieux Fort

Cape Moule à Chique

61°W

Saint Vincent Passage

A
B
C

4

CARIBBEAN SEA

80°W
60°W

Cocos
Islands

40°W

ATLANTIC

OCEAN

Lake
Maracaibo

River Orinoco

Mt. Roraima
2810m

GUIANA

HIGHLANDS

L l a n o s

A N D E S

Equator
0°

Cotopaxi
5896m
Chimborazo
6310m

River Negro

River Amazon

0°

Rocas
Island

Galapagos
Islands

Amazon Basin

River Amazon

S e l v a s

River Ucayali

River Madeira

River Tapajos

River Tacantins

Mato

Grosso

River São Francisco

B R A Z I L I A N

HIGHLANDS

3

PACIFIC

OCEAN

A N D E S

Lake
Titicaca

Lake
Poopo

River Pilcomayo

River Paraguay

20°S

20°S

Tropic of Capricorn

Atacama Desert

6908m
Ojos del
Salado

G r a n
C h a c o

Aconcagua
6960m

River Paraná

River Uruguay

Key

land height in metres
above sea level

more than
5000m

2000 – 5000m

1000 – 2000m

500 – 1000m

200 – 500m

less than
200 metres

land below sea level

▲ highest peaks with
heights in metres

⬭ lake

～ river

Juan
Fernández
Islands

P a m p a s

Rio de la Plata

ATLANTIC

OCEAN

N

R. Colorado

R. Negro

2

Chiloé
Island

Patagonia

A N D E S

Valdés
Peninsula

40°S

40°S

Falkland
Islands

Tierra
del Fuego

Cape Horn

South
Georgia

1

Scale One centimetre on the map
represents 350 kilometres
on the ground.

0   350   700   1050km

SOUTHERN OCEAN

60°S

80°W

60°S

A
B
C
D

80°W
60°W
40°W
20°W

60°S

Oblique Mercator Projection   © Oxford University Pr

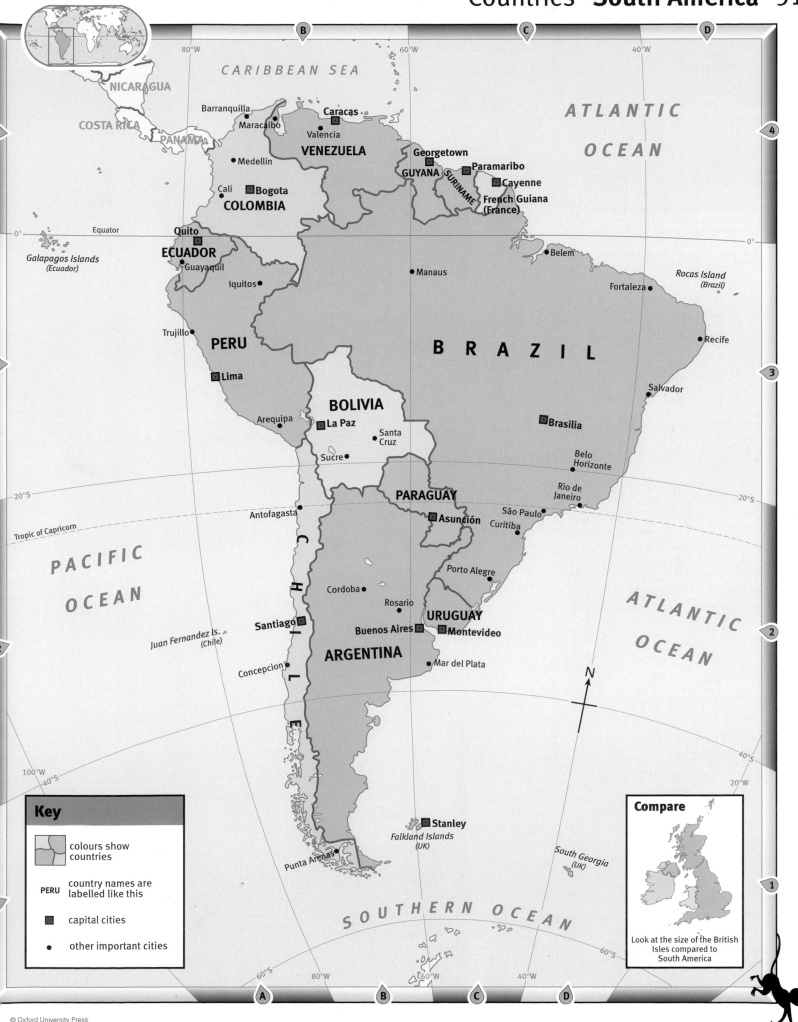

ATLANTIC OCEAN

CARIBBEAN SEA

NICARAGUA

COSTA RICA

PANAMA

Barranquilla
Maracaibo
Caracas
Valencia
VENEZUELA
Georgetown
Medellin
GUYANA
Paramaribo
SURINAME
Cayenne
Cali
Bogota
French Guiana
(France)
COLOMBIA

Equator

Galapagos Islands
(Ecuador)

Quito
ECUADOR
Guayaquil
Belem
Manaus
Rocas Island
(Brazil)
Iquitos
Fortaleza

PERU
B R A Z I L
Recife
Trujillo

Lima
Salvador

BOLIVIA
La Paz
Brasilia
Arequipa
Santa Cruz
Sucre
Belo Horizonte

Rio de Janeiro

PARAGUAY
São Paulo
Antofagasta
Asunción
Curitiba

Tropic of Capricorn

PACIFIC OCEAN

Porto Alegre

Cordoba
ATLANTIC OCEAN
Rosario
URUGUAY
Santiago
Buenos Aires
Montevideo
ARGENTINA
Concepcion
Mar del Plata

N

Juan Fernandez Is.
(Chile)

Stanley
Falkland Islands
(UK)
Punta Arenas
South Georgia
(UK)

SOUTHERN OCEAN

Compare

Key

colours show countries

PERU country names are labelled like this

■ capital cities

• other important cities

Look at the size of the British Isles compared to South America

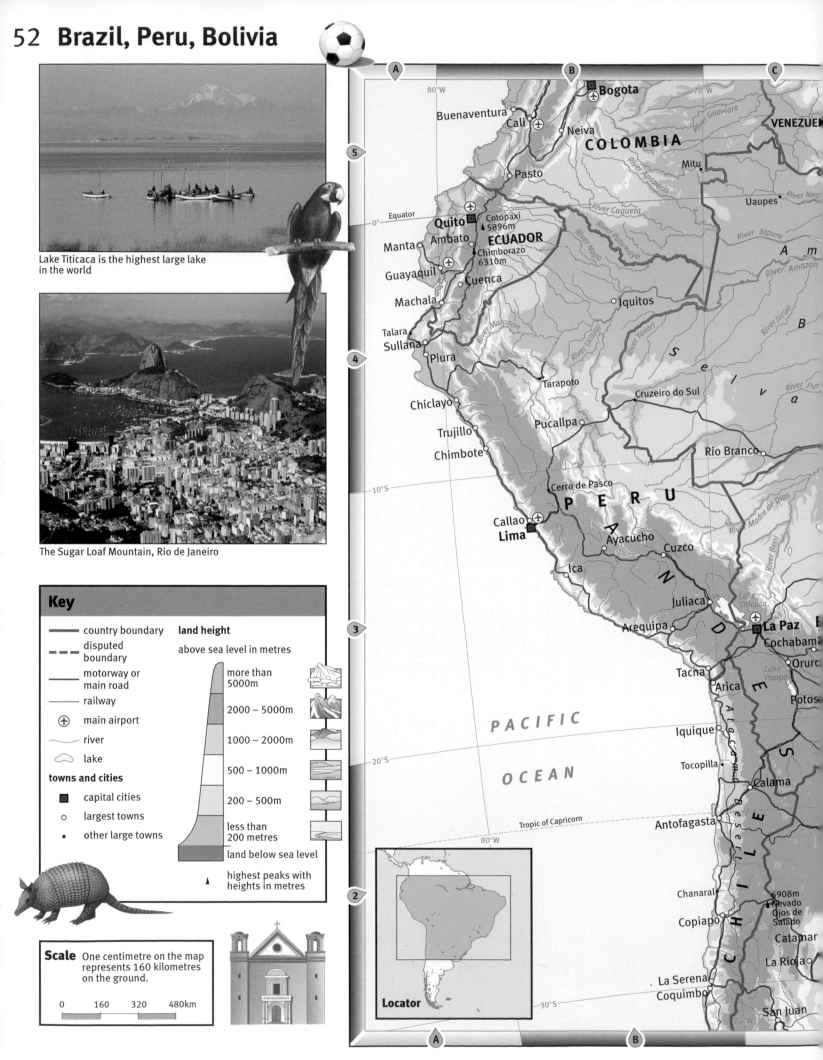

Lake Titicaca is the highest large lake in the world

The Sugar Loaf Mountain, Rio de Janeiro

Key

———	country boundary
– – –	disputed boundary
———	motorway or main road
——	railway
✈	main airport
∿	river
⬭	lake

towns and cities

▣	capital cities
○	largest towns
•	other large towns

land height

above sea level in metres

more than 5000m	
2000 – 5000m	
1000 – 2000m	
500 – 1000m	
200 – 500m	
less than 200 metres	
land below sea level	

▲ highest peaks with heights in metres

Scale One centimetre on the map represents 160 kilometres on the ground.

0   160   320   480km

Map labels

80°W
Bogota
70°W
Buenaventura
Cali
Neiva
COLOMBIA
Mitu
Pasto
River Apaporis
River Guaviare
VENEZUELA
River Neg
Uaupes
River Caqueta
Equator 0°
Quito
Cotopaxi 5896m
Manta
Ambato
ECUADOR
Chimborazo 6310m
River Putumayo
River Napo
River Amazon
River Japura
Am
Guayaquil
Cuenca
Machala
Iquitos
River Jurua
Talara
River Maranon
River Ucayali
River Yavari
Se
l
v
a
B
Sullana
Piura
Tarapoto
River Pur
Chiclayo
Cruzeiro do Sul
Trujillo
Pucallpa
Chimbote
Rio Branco
Cerro de Pasco
10°S
P E R U
River Madre de Dios
Callao
Lima
Ayacucho
Cuzco
A
River Beni
Ica
N
Juliaca
Lake Titicaca
Arequipa
D
La Paz
Cochabam
Tacna
E
Lake Poopo
Oruro
Arica
S
Potos
PACIFIC
Iquique
Atacama Desert
Tocopilla
20°S
OCEAN
Calama
Tropic of Capricorn
Antofagasta
80°W
C H I L E
Chanaral
6908m Nevado Ojos de Salado
Copiapo
Catamar
La Rioja
La Serena
Coquimbo
30°S
70°W
San Juan

Locator

© Oxford University Press
Transverse Mercator Projection

GUYANA
SURINAME
French Guiana (France)

Boa Vista

Serra Tumucumaque

ATLANTIC

OCEAN

Macapa

Barcelos

Balbina Reservoir

River Branco

Mouths of the Amazon

Ilha de Marajo

Braganca

Equator 0°

Belem
São Luis

Manaus
Santarem
Cameta

River Amazon

Parnaiba

Manacapuru

A z o n
River Madeira

Coari

Itaituba
River Iriri

Altamira

Tucurui

Bacabal
Codo
Caxias
Sobral
Fortaleza

s i n

River Tapajos

R. Xingu

Maraba

Imperatriz
Teresina
Mossoro

Porto Velho

River Aripuana

River Teles Pires

Araguaina

Barra do Corda

Natal
Juazeiro do Norte
Campina Grande
Joao Pessoa

Ariquemes

River Juruena

River Xingu

River Araguaia

River Parnaiba

River Tocantins

Petrolina
Caruaru
Recife

B R A Z I L

Barreiras

Maceio 10°S

River Guapore

Feira de Santana
Aracaju

rinidad

Mato Grosso

Diamantina

River São Francisco
Chapada

Alagoinhas
Salvador

Cuiaba

B R A Z I L I A N

Vitoria da Conquista
Jequie
Ilheus

LIVIA

Caceres

Rondonopolis

Anapolis
Brasilia

Santa Cruz

Goiania

Montes Claros

cre

Corumba

Rio Verde

H I G H L A N D S

River Jequitinhonha

Teofilo Otoni

River Paranaiba

Mount Itambe 2033m

Governador Valadares

Sa. de Maracaju

Uberlandia

Linhares

Tarija

Caratinga

River Paraguay

Uberaba

Ribeirao Preto
Belo Horizonte
Vitoria

São Jose do Rio Preto

Campo Grande

Barbacena

Araraquara

Juiz de Fora

Pedro Juan Caballero
Dourados
Bauru

Campos

n Salvador
e Jujuy

Gran Chaco

River Pilcomayo

Maringa
Campinas
Nova Iguacu
Rio de Janeiro

alta

São Paulo
Santo Andre

PARAGUAY

Ponta Grossa

Santos

River Bermejo

Asuncion

Foz do Iguacu

Curitiba

Tropic of Capricorn

River Parana

Formosa

Paranagua

ATLANTIC

San Miguel de Tucuman
Resistencia

Posadas

Itajai

OCEAN

GENTINA
Corrientes

Florianopolis

Santiago del Estero

River Uruguay

Passo Fundo

River Salado

Santa Maria
Caxias do Sul

Uruguaiana

s Grande

Concordia

Porto Alegre

River Parana

Santa Fe
Parana
URUGUAY

Lagoa dos Patos 50°W

ordoba

Pelotas

Rio Grande

© Oxford University Press

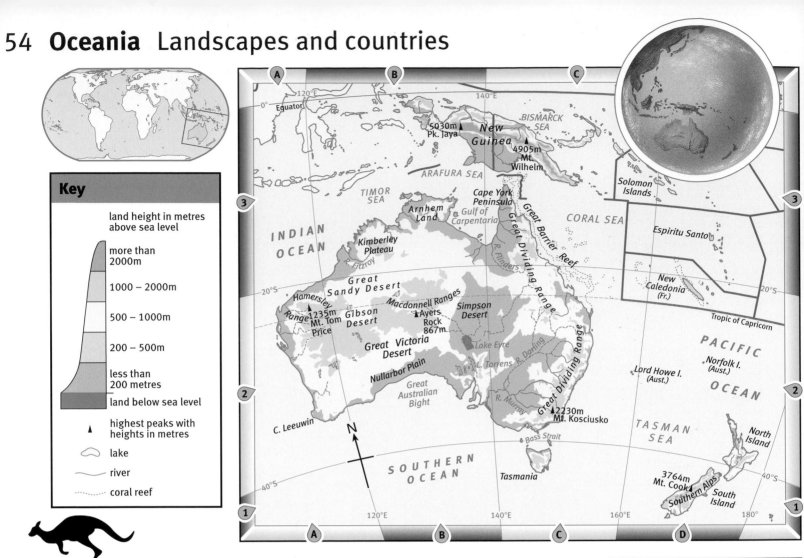

Key

land height in metres above sea level

more than 2000m

1000 – 2000m

500 – 1000m

200 – 500m

less than 200 metres

land below sea level

▲ highest peaks with heights in metres

⌒ lake

⌒ river

⋯ coral reef

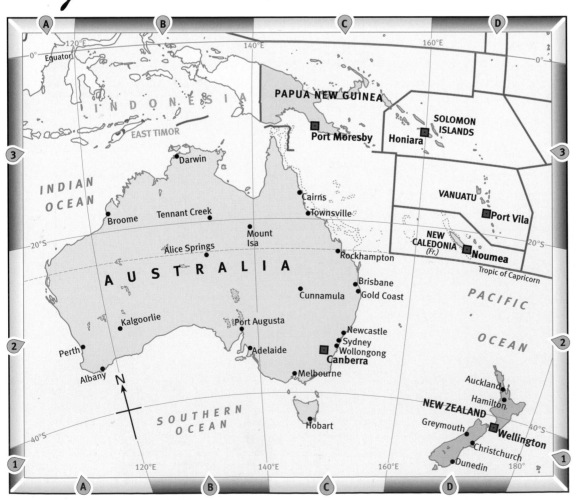

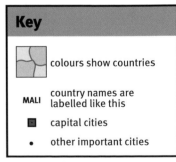

Scale One centimetre on these maps represents 450 kilometres on the ground.

| 0 | 450 | 900 | 1350km |

Key

colours show countries

MALI country names are labelled like this

▪ capital cities

• other important cities

Compare

Look at the size of the British Isles compared to Oceania

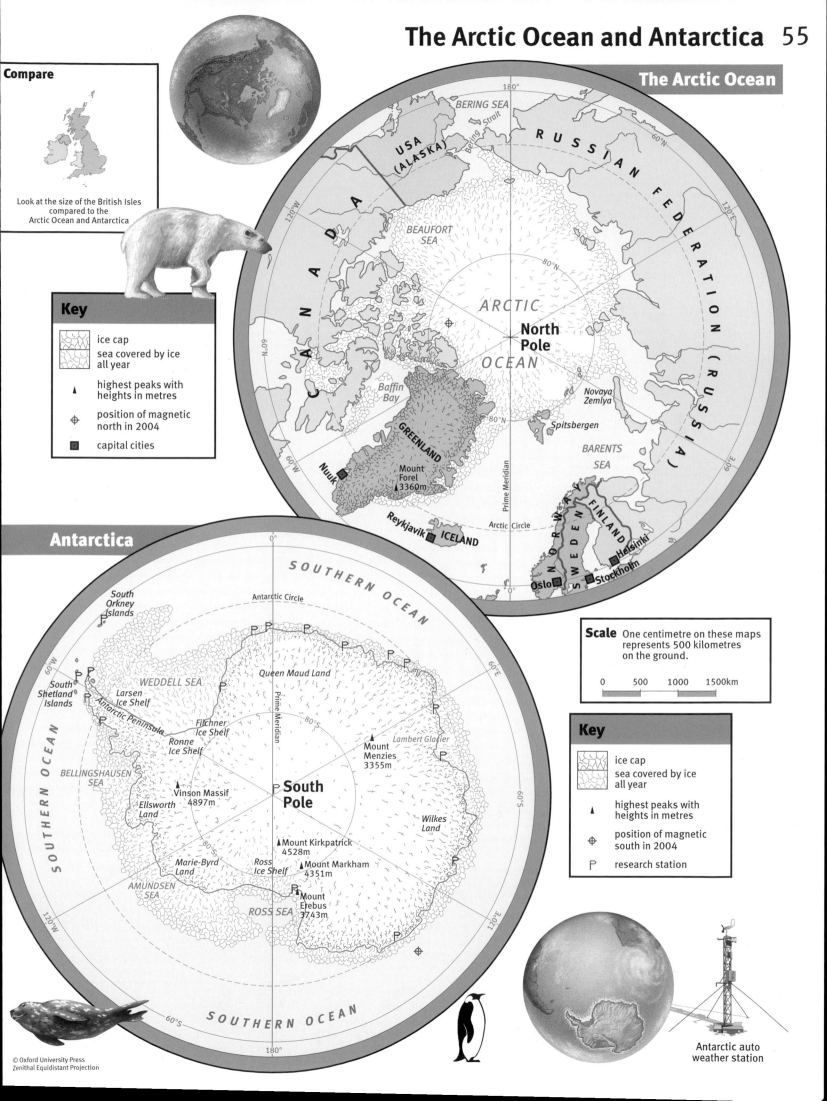

Compare

Look at the size of the British Isles compared to the Arctic Ocean and Antarctica

Key

ice cap

sea covered by ice all year

▲ highest peaks with heights in metres

⊕ position of magnetic north in 2004

■ capital cities

The Arctic Ocean map labels:

BERING SEA
Bering Strait
USA (ALASKA)
RUSSIAN FEDERATION (RUSSIA)
60°N
120°E
CANADA
BEAUFORT SEA
80°N
ARCTIC OCEAN
North Pole
Baffin Bay
80°N
Novaya Zemlya
Spitsbergen
GREENLAND
Mount Forel ▲3360m
BARENTS SEA
Prime Meridian
Arctic Circle
Nuuk ■
Reykjavik ■ ICELAND
NORWAY
SWEDEN
FINLAND
Helsinki ■
Oslo ■
Stockholm ■
120°W
60°W
60°N
0°
60°E

Antarctica

Antarctica map labels:

SOUTHERN OCEAN
0°
Antarctic Circle
South Orkney Islands
South Shetland Islands
Antarctic Peninsula
Larsen Ice Shelf
WEDDELL SEA
Queen Maud Land
Filchner Ice Shelf
Ronne Ice Shelf
Prime Meridian
80°S
Lambert Glacier
Mount Menzies 3355m
BELLINGSHAUSEN SEA
Vinson Massif 4897m
Ellsworth Land
South Pole
Wilkes Land
Marie-Byrd Land
Mount Kirkpatrick 4528m
Ross Ice Shelf
Mount Markham 4351m
AMUNDSEN SEA
Mount Erebus 3743m
ROSS SEA
SOUTHERN OCEAN
60°W
120°W
60°S
180°
60°E
120°E
50°S
SOUTHERN OCEAN

Scale One centimetre on these maps represents 500 kilometres on the ground.

0   500   1000   1500km

Key

ice cap

sea covered by ice all year

▲ highest peaks with heights in metres

⊕ position of magnetic south in 2004

P research station

Antarctic auto weather station

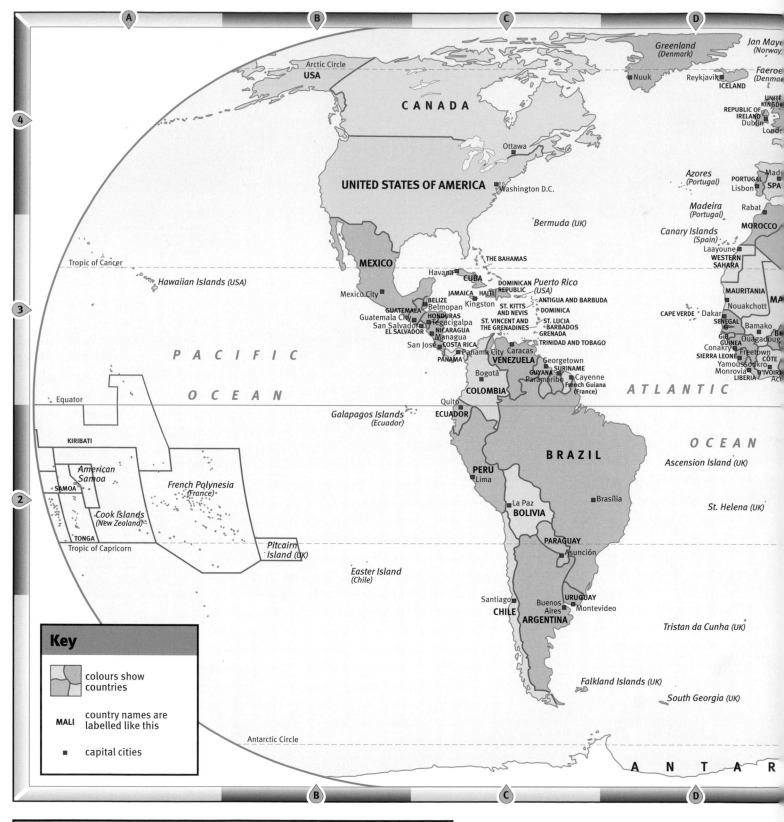

Arctic Circle
USA
CANADA
Greenland (Denmark)
Jan Maye (Norway)
Faeroe (Denma
Nuuk
Reykjavik
ICELAND
UNITE KINGD
REPUBLIC OF IRELAND
Dublin
Lond

Ottawa

UNITED STATES OF AMERICA
Washington D.C.

Azores (Portugal)
PORTUGAL
Lisbon
SPA
Mad

Bermuda (UK)
Rabat
MOROCCO

Tropic of Cancer
Hawaiian Islands (USA)
MEXICO
THE BAHAMAS
Mexico City
Havana
CUBA
DOMINICAN REPUBLIC
Puerto Rico (USA)
Canary Islands (Spain)
Laayoune
WESTERN SAHARA

JAMAICA
HAITI
ANTIGUA AND BARBUDA
MAURITANIA
Nouakchott
MA

BELIZE
Belmopan
Kingston
ST. KITTS AND NEVIS
DOMINICA
GUATEMALA
Guatemala City
HONDURAS
Tegucigalpa
ST. VINCENT AND THE GRENADINES
ST. LUCIA
BARBADOS
CAPE VERDE
Dakar
SENEGAL
G
Bamako
San Salvador
EL SALVADOR
NICARAGUA
Managua
GRENADA
GIB
GUINEA
Ouagadoug

PACIFIC
San José
COSTA RICA
PANAMA
Panama City
Caracas
TRINIDAD AND TOBAGO
Conakry
SIERRA LEONE
Freetown
CÔTE
D'IVOIR

OCEAN
Bogotá
VENEZUELA
Georgetown
SURINAME
GUYANA
Paramaribo
Cayenne
French Guiana (France)
ATLANTIC
Yamoussoukro
Monrovia
LIBERIA
Ac

COLOMBIA
Quito
ECUADOR
Galapagos Islands (Ecuador)

Equator

KIRIBATI
BRAZIL
OCEAN
Ascension Island (UK)

American Samoa
PERU
Lima
SAMOA
French Polynesia (France)
La Paz
Brasília
St. Helena (UK)

Cook Islands (New Zealand)
BOLIVIA
TONGA
PARAGUAY
Asunción
Tropic of Capricorn
Pitcairn Island (UK)
Easter Island (Chile)
Santiago
Buenos Aires
URUGUAY
Montevideo
Tristan da Cunha (UK)
CHILE
ARGENTINA

Falkland Islands (UK)
South Georgia (UK)

Antarctic Circle
ANTAR

Key

colours show countries

MALI country names are labelled like this

■ capital cities

© Oxford University Press   Eckert IV Projection

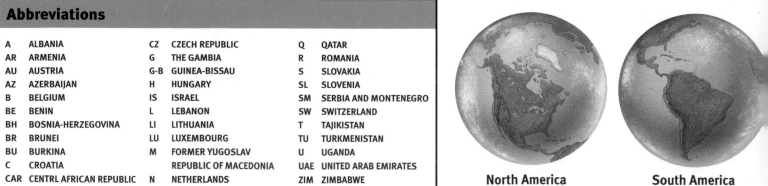

Abbreviations

A	ALBANIA	CZ	CZECH REPUBLIC	Q	QATAR
AR	ARMENIA	G	THE GAMBIA	R	ROMANIA
AU	AUSTRIA	G-B	GUINEA-BISSAU	S	SLOVAKIA
AZ	AZERBAIJAN	H	HUNGARY	SL	SLOVENIA
B	BELGIUM	IS	ISRAEL	SM	SERBIA AND MONTENEGRO
BE	BENIN	L	LEBANON	SW	SWITZERLAND
BH	BOSNIA-HERZEGOVINA	LI	LITHUANIA	T	TAJIKISTAN
BR	BRUNEI	LU	LUXEMBOURG	TU	TURKMENISTAN
BU	BURKINA	M	FORMER YUGOSLAV	U	UGANDA
C	CROATIA		REPUBLIC OF MACEDONIA	UAE	UNITED ARAB EMIRATES
CAR	CENTRL AFRICAN REPUBLIC	N	NETHERLANDS	ZIM	ZIMBABWE

North America

South America

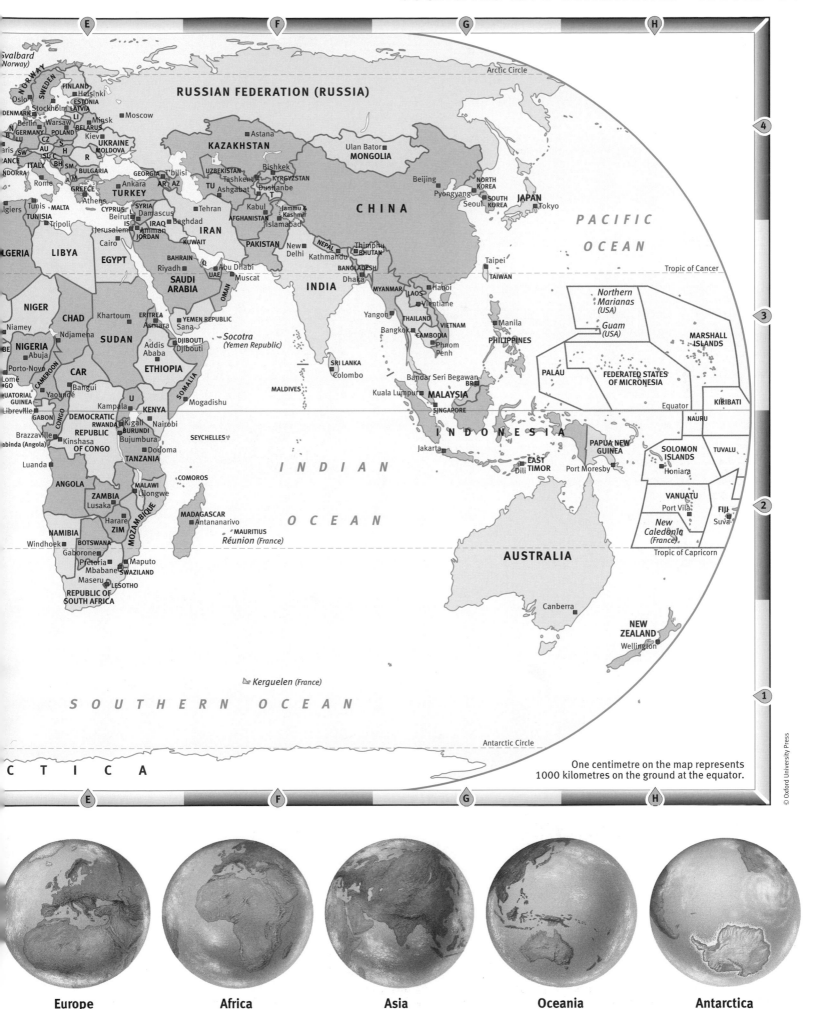

One centimetre on the map represents
1000 kilometres on the ground at the equator.

© Oxford University Press

Europe **Africa** **Asia** **Oceania** **Antarctica**

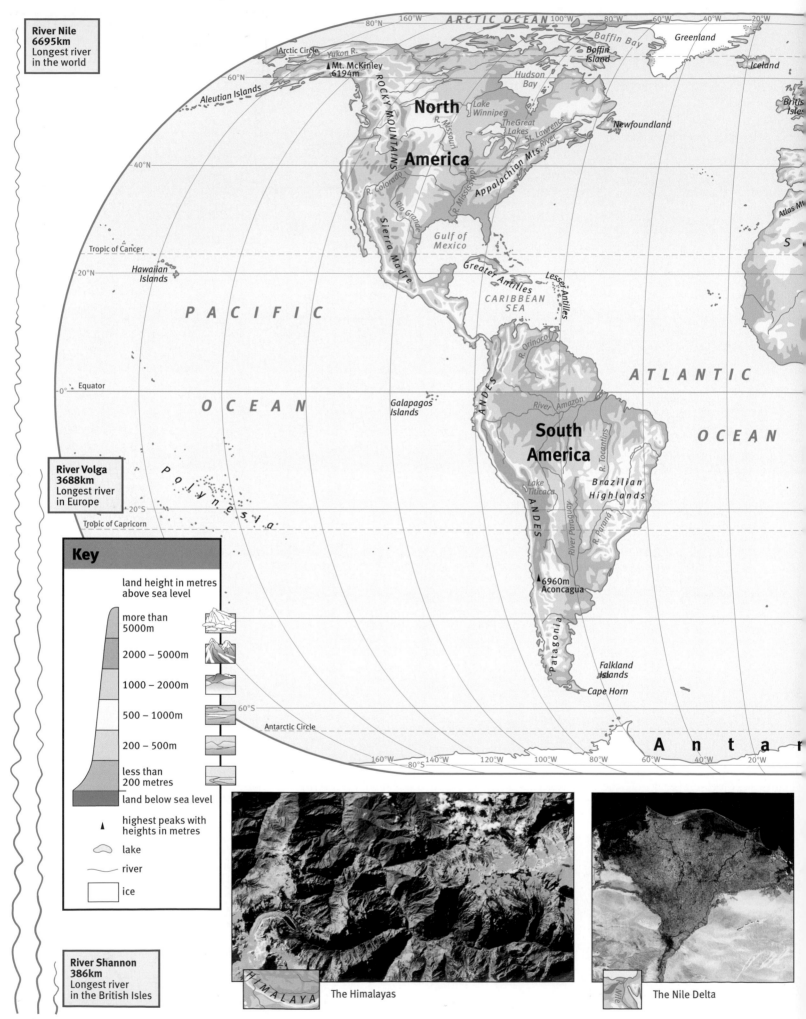

River Nile
6695km
Longest river
in the world

River Volga
3688km
Longest river
in Europe

River Shannon
386km
Longest river
in the British Isles

ARCTIC OCEAN

Baffin Bay
Greenland
Baffin
Island
Iceland

Arctic Circle
Yukon R.
▲Mt. McKinley
6194m
Hudson
Bay
Lake
Winnipeg
Newfoundland
Britis
Isles

North

ROCKY MOUNTAINS
R.
Missouri
TheGreat
Lakes
St. Lawrence

America

R. Colorado
R. Mississippi
Appalachian Mts. River.
Atlas Mt
S

Tropic of Cancer
Sierra Madre
Rio Grande
Gulf of
Mexico
Hawaiian
Islands
Greater Antilles
Lesser Antilles
CARIBBEAN
SEA

PACIFIC
Galapagos
Islands
R. Orinoco

Equator

OCEAN
ANDES
River Amazon
ATLANTIC

South
America
R. Tocantins
OCEAN

Polynesia
Lake
Titicaca
Brazilian
Highlands

Tropic of Capricorn
ANDES
River Paraguay
R. Parana

▲6960m
Aconcagua

Key

land height in metres
above sea level

more than
5000m

2000 – 5000m

1000 – 2000m

500 – 1000m

200 – 500m

less than
200 metres

land below sea level

▲ highest peaks with
heights in metres

lake

river

ice

Patagonia
Falkland
Islands
Cape Horn

60°S
Antarctic Circle

Antar

HIMALAYA The Himalayas

Nile The Nile Delta

© Oxford University Press
Eckert IV Projection

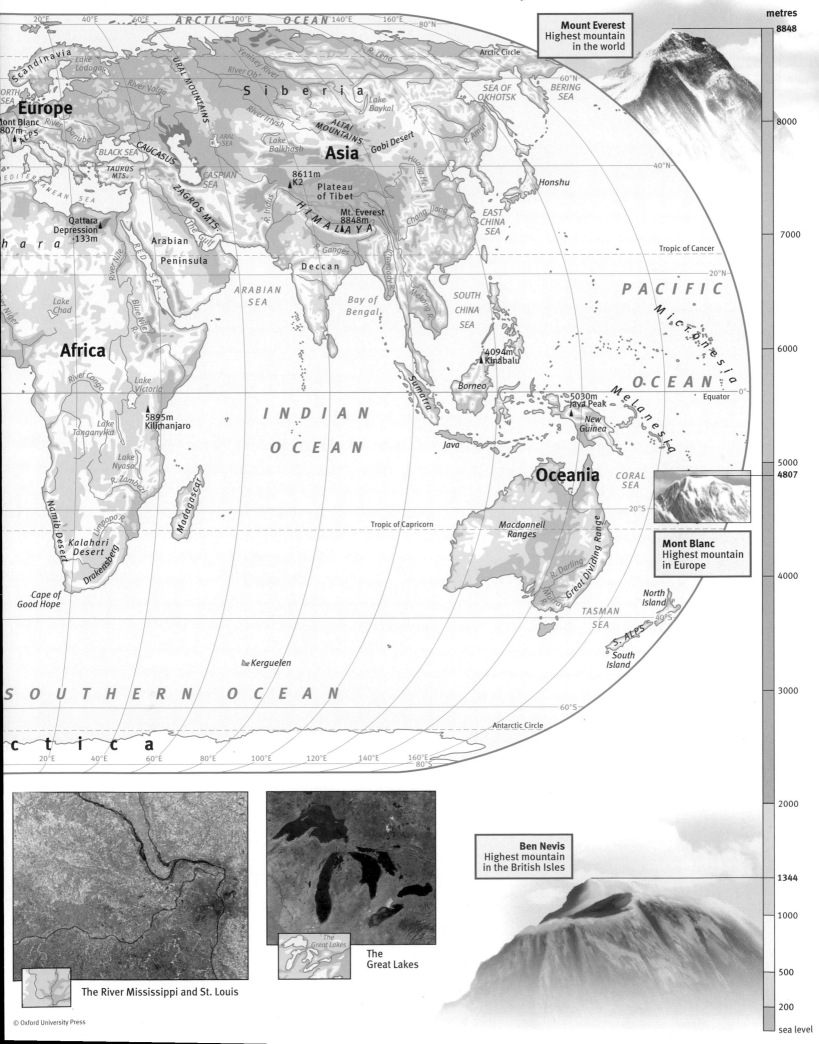

Mount Everest
Highest mountain in the world

Mont Blanc
Highest mountain in Europe

Ben Nevis
Highest mountain in the British Isles

metres
8848
8000
7000
6000
5000
4807
4000
3000
2000
1344
1000
500
200
sea level

The River Mississippi and St. Louis

The Great Lakes

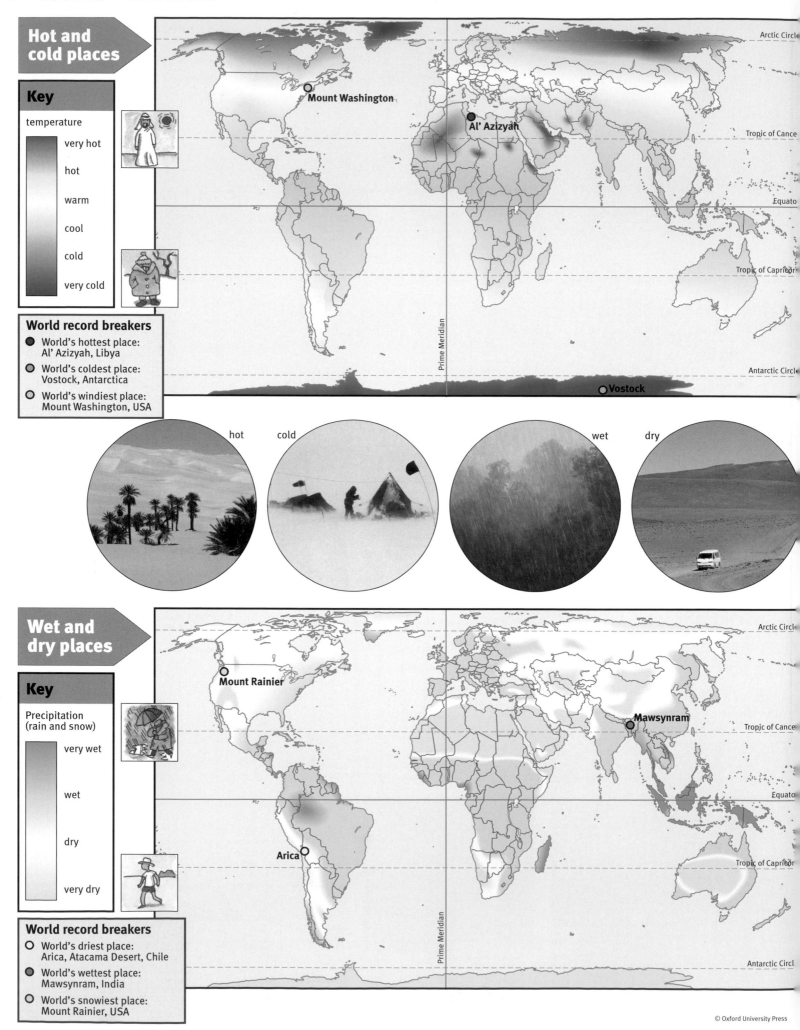

Hot and cold places

Key

temperature

- very hot
- hot
- warm
- cool
- cold
- very cold

World record breakers

- ● World's hottest place:
  Al' Azizyah, Libya
- ◑ World's coldest place:
  Vostock, Antarctica
- ○ World's windiest place:
  Mount Washington, USA

Mount Washington

Al' Azizyah

Vostock

Arctic Circle
Tropic of Cancer
Equator
Tropic of Capricorn
Antarctic Circle
Prime Meridian

hot cold wet dry

Wet and dry places

Key

Precipitation
(rain and snow)

- very wet
- wet
- dry
- very dry

World record breakers

- ○ World's driest place:
  Arica, Atacama Desert, Chile
- ● World's wettest place:
  Mawsynram, India
- ◑ World's snowiest place:
  Mount Rainier, USA

Mount Rainier

Mawsynram

Arica

Arctic Circle
Tropic of Cancer
Equator
Tropic of Capricorn
Antarctic Circle
Prime Meridian

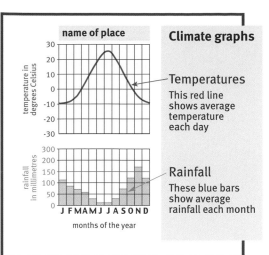

name of place · **Climate graphs**

Temperatures
This red line shows average temperature each day

Rainfall
These blue bars show average rainfall each month

months of the year

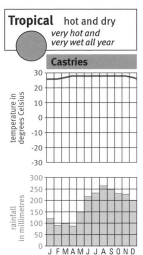

Tropical hot and dry
very hot and very wet all year

Castries

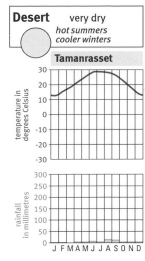

Desert very dry
hot summers cooler winters

Tamanrasset

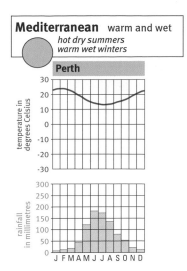

Mediterranean warm and wet
hot dry summers warm wet winters

Perth

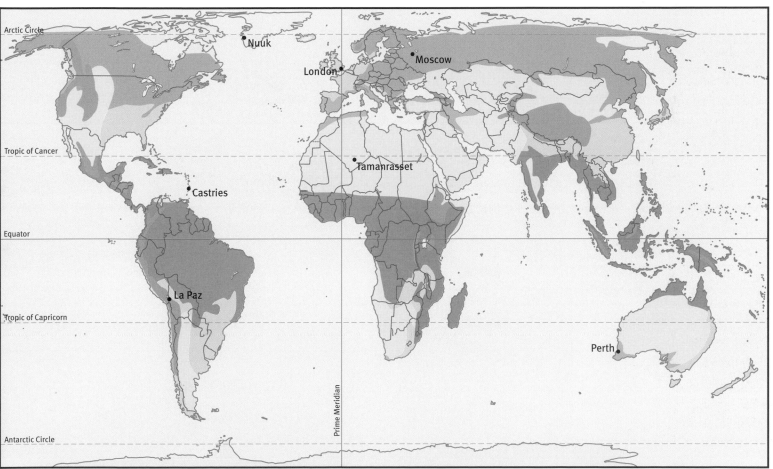

Arctic Circle
Nuuk
Moscow
London
Tropic of Cancer
Tamanrasset
Castries
Equator
La Paz
Tropic of Capricorn
Perth
Prime Meridian
Antarctic Circle

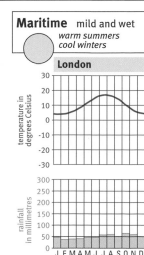

Maritime mild and wet
warm summers cool winters

London

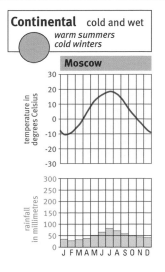

Continental cold and wet
warm summers cold winters

Moscow

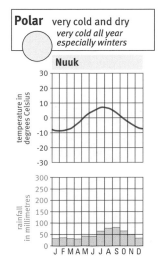

Polar very cold and dry
very cold all year especially winters

Nuuk

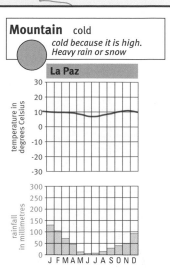

Mountain cold
cold because it is high. Heavy rain or snow

La Paz

**There are about
6 300 000 000
people in the world.**

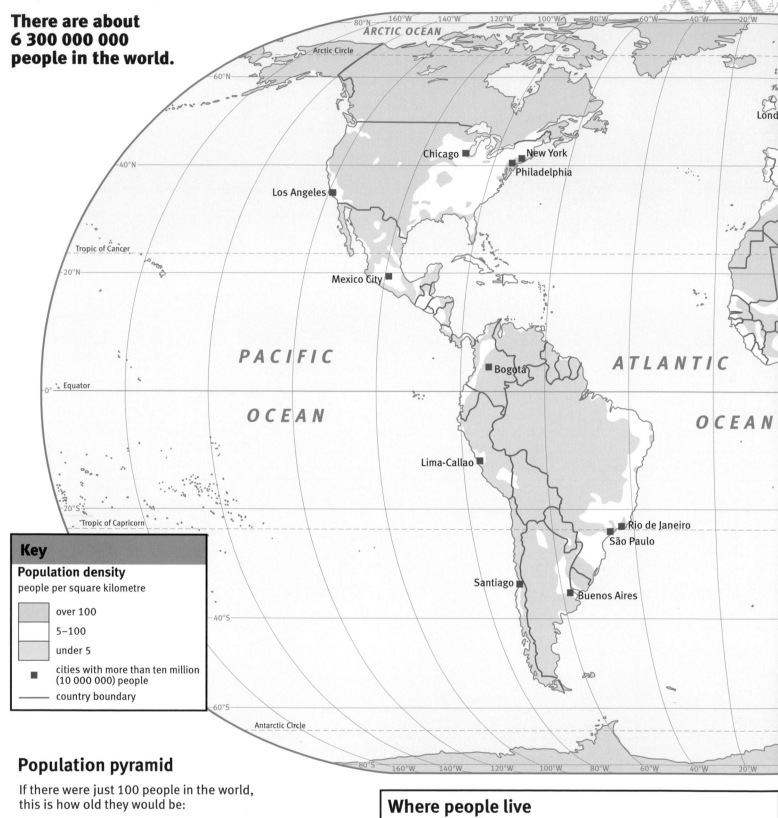

ARCTIC OCEAN

Arctic Circle

Chicago

New York
Philadelphia

Los Angeles

Tropic of Cancer

Mexico City

PACIFIC

Bogotá

ATLANTIC

Equator

OCEAN

OCEAN

Lima-Callao

Rio de Janeiro
São Paulo

Santiago

Buenos Aires

Tropic of Capricorn

Antarctic Circle

London

Key

Population density

people per square kilometre

	over 100
	5–100
	under 5
■	cities with more than ten million (10 000 000) people
——	country boundary

Population pyramid

If there were just 100 people in the world,
this is how old they would be:

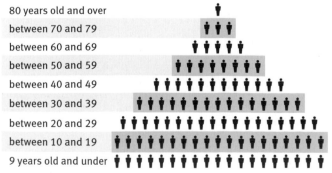

80 years old and over	
between 70 and 79	
between 60 and 69	
between 50 and 59	
between 40 and 49	
between 30 and 39	
between 20 and 29	
between 10 and 19	
9 years old and under	

Where people live

If there were just
100 people in
the world, this
is where they
would live:

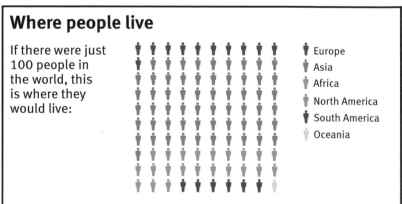

- Europe
- Asia
- Africa
- North America
- South America
- Oceania

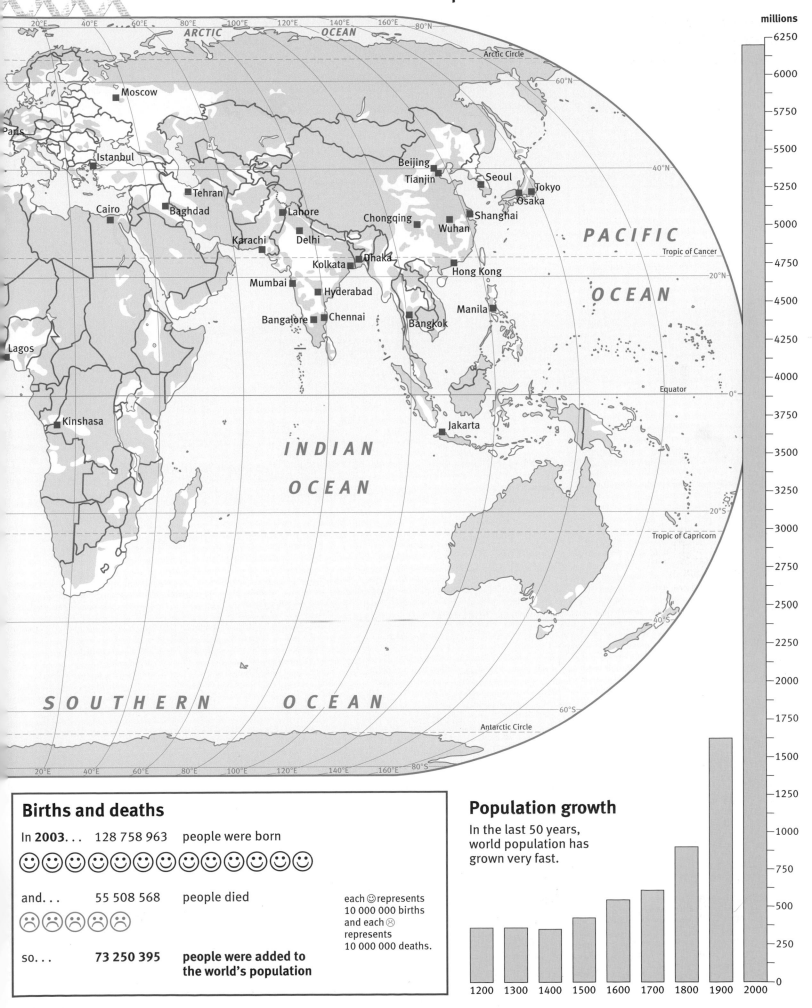

millions
- 6250
- 6000
- 5750
- 5500
- 5250
- 5000
- 4750
- 4500
- 4250
- 4000
- 3750
- 3500
- 3250
- 3000
- 2750
- 2500
- 2250
- 2000
- 1750
- 1500
- 1250
- 1000
- 750
- 500
- 250
- 0

Births and deaths

In **2003**. . . 128 758 963 people were born

☺☺☺☺☺☺☺☺☺☺☺☺☺

and. . . 55 508 568 people died

☹☹☹☹☹

so. . . **73 250 395** **people were added to the world's population**

each ☺ represents
10 000 000 births
and each ☹
represents
10 000 000 deaths.

Population growth

In the last 50 years, world population has grown very fast.

1200 1300 1400 1500 1600 1700 1800 1900 2000

© Oxford University Press

tropical forest

deciduous forest

coniferous forest

ARCTIC OCEAN

Arctic Circle

80°N   160°W   140°W   120°W   100°W   80°W   60°W   40°W   20°W

60°N

40°N

Tropic of Cancer

20°N

PACIFIC

0° Equator

OCEAN

ATLANTIC

OCEAN

20°S

Tropic of Capricorn

40°S

60°S

Antarctic Circle

80°S   160°W   140°W   120°W   100°W   80°W   60°W   40°W   20°W

Key

coniferous forest
trees have leaves all year

deciduous forest
trees drop their leaves in winter

tropical forest
tall trees growing close together

savannah
tall trees and scattered trees

temperate grassland
prairies, steppes, pampas and veld

semi desert
short grass and small dry bushes

desert
sand and stones with few plants

tundra
moss and bog with some short trees

ice
no plants

mountains
thin soils and steep slopes

desert

semi desert

savannah

temperate grassland

ARCTIC OCEAN

Arctic Circle

60°N

40°N

PACIFIC

Tropic of Cancer

20°N

OCEAN

Equator 0°

INDIAN

OCEAN

20°S

Tropic of Capricorn

40°S

© Oxford University Press

SOUTHERN OCEAN

60°S

Antarctic Circle

80°S

20°E 40°E 60°E 80°E 100°E 120°E 140°E 160°E 80°S

tundra

mountains

ice

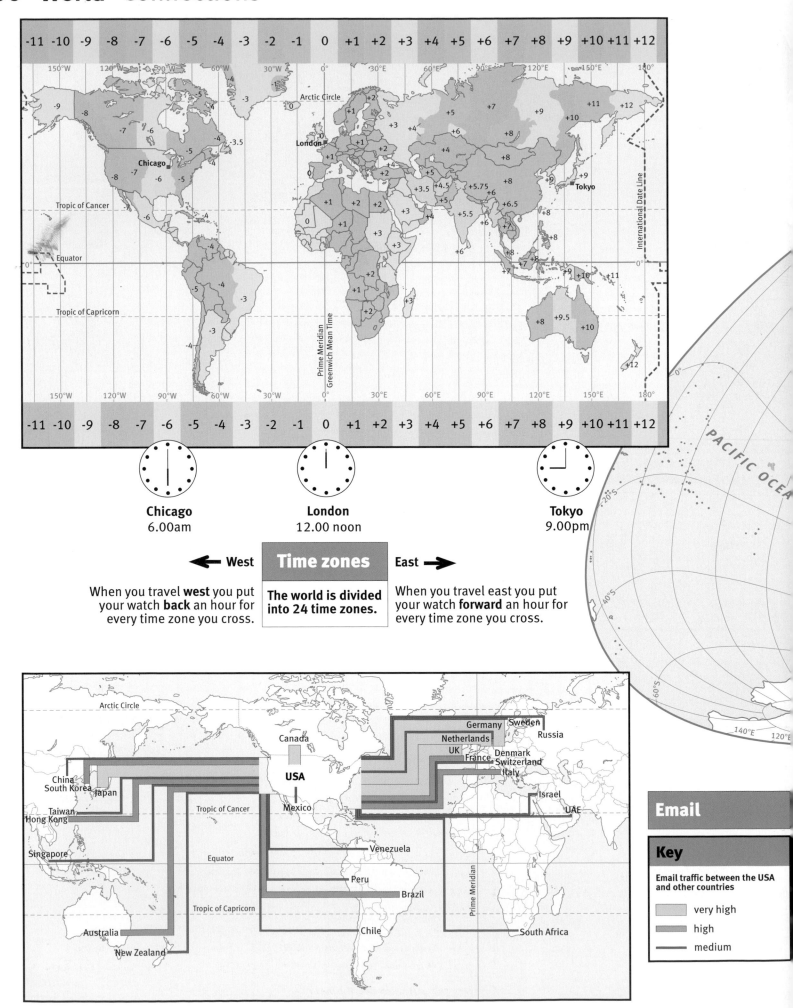

Chicago
6.00am

London
12.00 noon

Tokyo
9.00pm

← **West**

Time zones

East →

When you travel **west** you put your watch **back** an hour for every time zone you cross.

The world is divided into 24 time zones.

When you travel east you put your watch **forward** an hour for every time zone you cross.

Email

Key

Email traffic between the USA and other countries

very high

high

medium

© Oxford University Press
Gall Projection

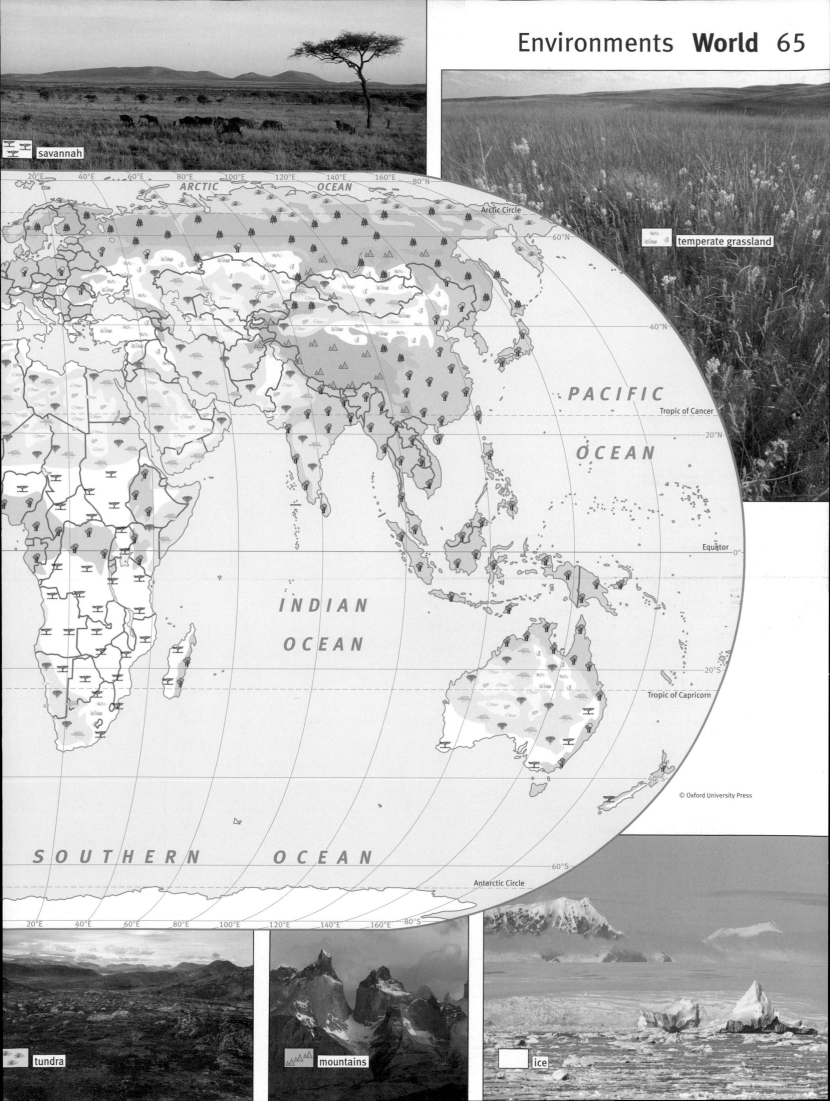

savannah

temperate grassland

ARCTIC OCEAN

Arctic Circle

60°N

40°N

PACIFIC

Tropic of Cancer

20°N

OCEAN

Equator 0°

INDIAN

OCEAN

20°S

Tropic of Capricorn

40°S

© Oxford University Press

SOUTHERN OCEAN

60°S

Antarctic Circle

80°S

20°E 40°E 60°E 80°E 100°E 120°E 140°E 160°E

80°N

tundra

mountains

ice

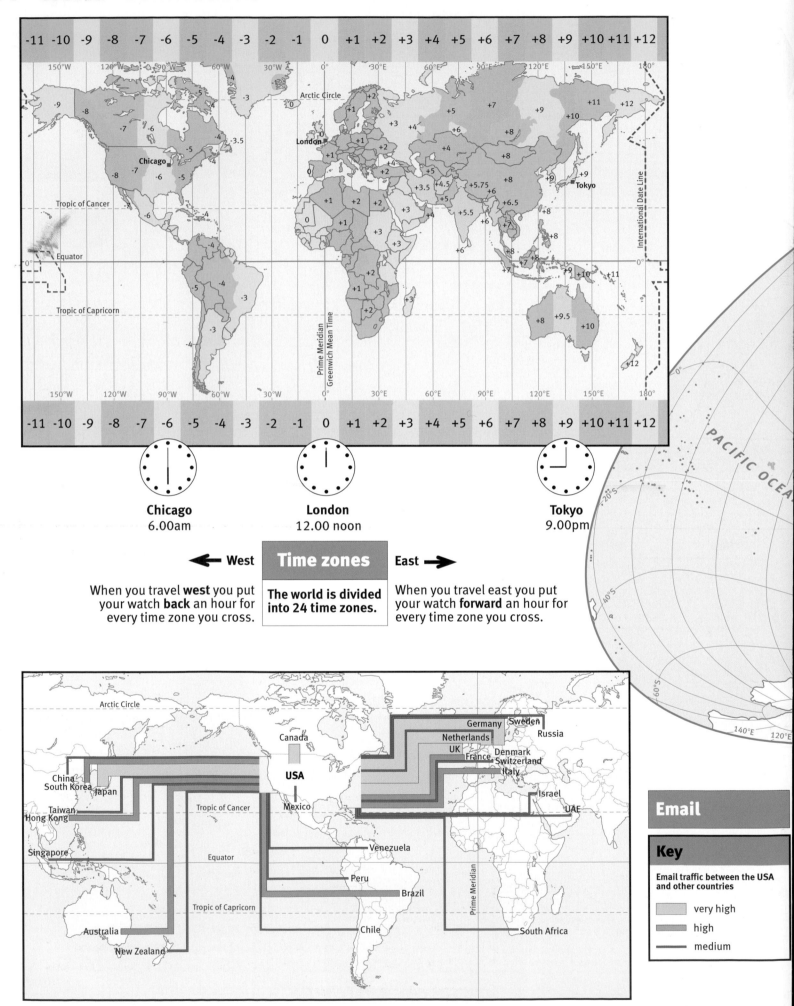

| -11 | -10 | -9 | -8 | -7 | -6 | -5 | -4 | -3 | -2 | -1 | 0 | +1 | +2 | +3 | +4 | +5 | +6 | +7 | +8 | +9 | +10 | +11 | +12 |

Chicago
6.00am

London
12.00 noon

Tokyo
9.00pm

← **West**

Time zones

East →

When you travel **west** you put your watch **back** an hour for every time zone you cross.

The world is divided into 24 time zones.

When you travel east you put your watch **forward** an hour for every time zone you cross.

Email

Key

Email traffic between the USA and other countries

very high

high

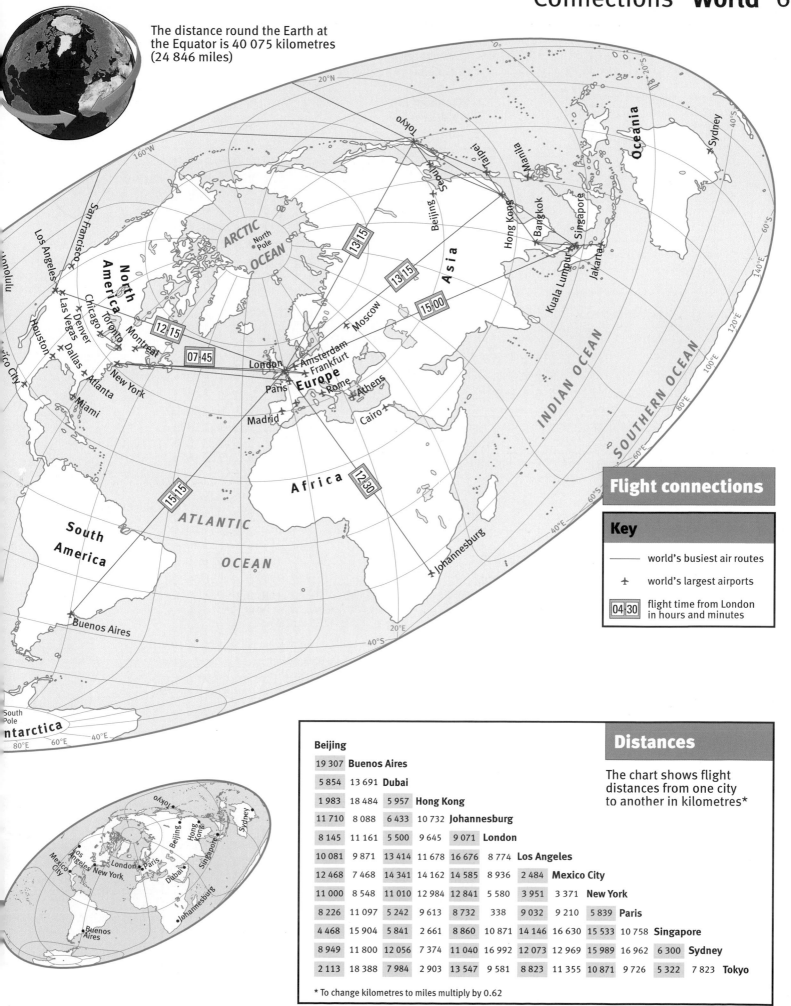

The distance round the Earth at the Equator is 40 075 kilometres (24 846 miles)

Flight connections

Key

———	world's busiest air routes
✈	world's largest airports
04 30	flight time from London in hours and minutes

Distances

The chart shows flight distances from one city to another in kilometres*

Beijing											
19 307	**Buenos Aires**										
5 854	13 691	**Dubai**									
1 983	18 484	5 957	**Hong Kong**								
11 710	8 088	6 433	10 732	**Johannesburg**							
8 145	11 161	5 500	9 645	9 071	**London**						
10 081	9 871	13 414	11 678	16 676	8 774	**Los Angeles**					
12 468	7 468	14 341	14 162	14 585	8 936	2 484	**Mexico City**				
11 000	8 548	11 010	12 984	12 841	5 580	3 951	3 371	**New York**			
8 226	11 097	5 242	9 613	8 732	338	9 032	9 210	5 839	**Paris**		
4 468	15 904	5 841	2 661	8 860	10 871	14 146	16 630	15 533	10 758	**Singapore**	
8 949	11 800	12 056	7 374	11 040	16 992	12 073	12 969	15 989	16 962	6 300	**Sydney**
2 113	18 388	7 984	2 903	13 547	9 581	8 823	11 355	10 871	9 726	5 322	7 823	**Tokyo**

* To change kilometres to miles multiply by 0.62

Oxford University Press
Oblique Aitoff Projection

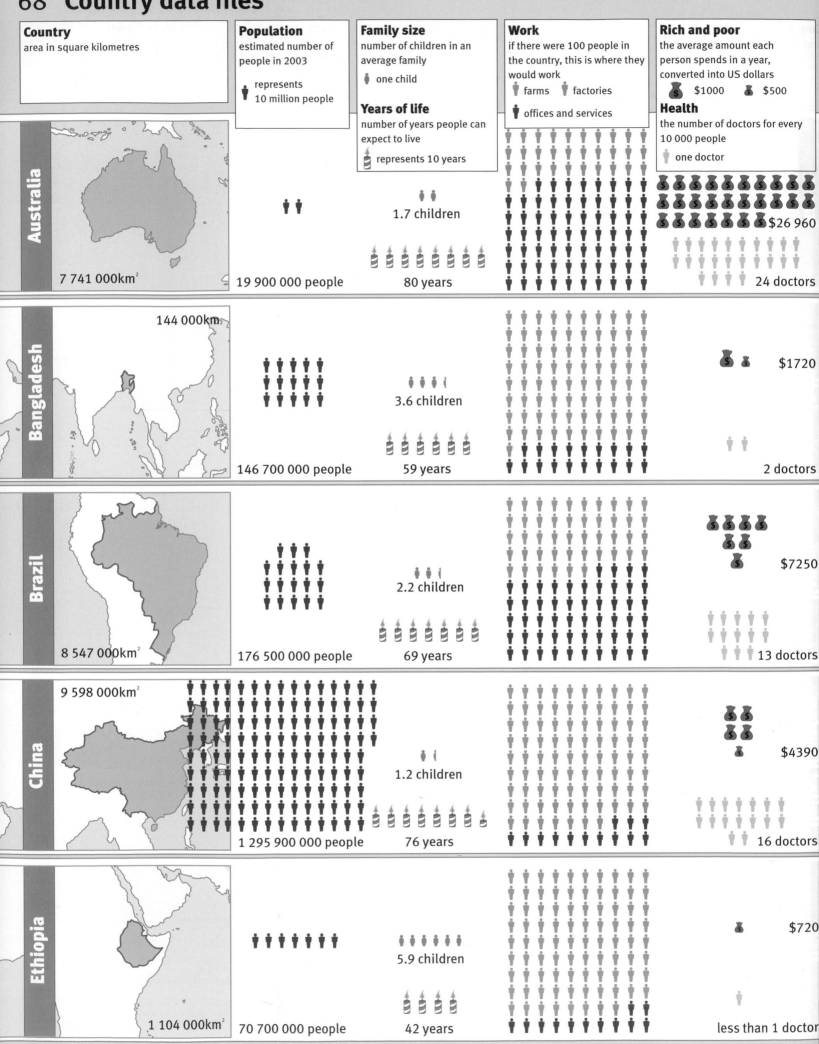

Country area in square kilometres	Population estimated number of people in 2003 represents 10 million people	Family size number of children in an average family one child Years of life number of years people can expect to live represents 10 years	Work if there were 100 people in the country, this is where they would work farms factories offices and services	Rich and poor the average amount each person spends in a year, converted into US dollars $1000 $500 Health the number of doctors for every 10 000 people one doctor
Australia 7 741 000km²	19 900 000 people	1.7 children 80 years		$26 960 24 doctors
Bangladesh 144 000km²	146 700 000 people	3.6 children 59 years		$1720 2 doctors
Brazil 8 547 000km²	176 500 000 people	2.2 children 69 years		$7250 13 doctors
China 9 598 000km²	1 295 900 000 people	1.2 children 76 years		$4390 16 doctors
Ethiopia 1 104 000km²	70 700 000 people	5.9 children 42 years		$720 less than 1 doctor

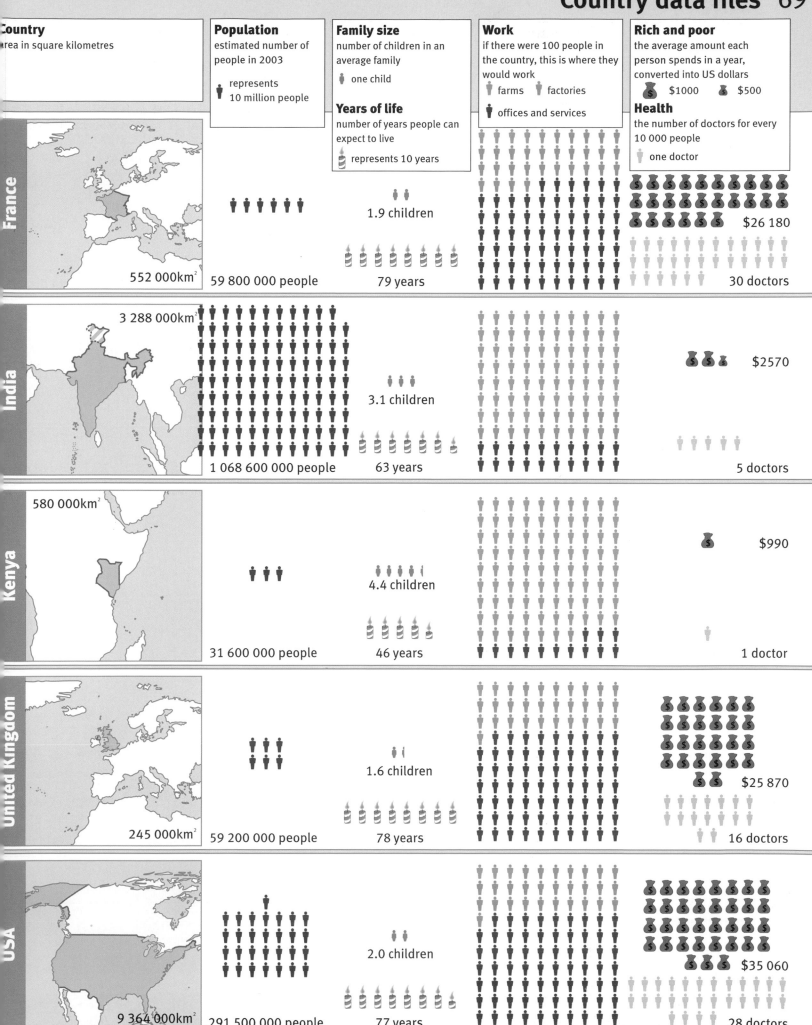

Country
area in square kilometres

Population
estimated number of people in 2003

represents 10 million people

Family size
number of children in an average family

one child

Years of life
number of years people can expect to live

represents 10 years

Work
if there were 100 people in the country, this is where they would work

farms   factories

offices and services

Rich and poor
the average amount each person spends in a year, converted into US dollars

$1000 $500

Health
the number of doctors for every 10 000 people

one doctor

France
552 000km²   59 800 000 people   1.9 children   79 years $26 180   30 doctors

India
3 288 000km²   1 068 600 000 people   3.1 children   63 years $2570   5 doctors

Kenya
580 000km²   31 600 000 people   4.4 children   46 years $990   1 doctor

United Kingdom
245 000km²   59 200 000 people   1.6 children   78 years $25 870   16 doctors

USA
9 364 000km²   291 500 000 people   2.0 children   77 years $35 060   28 doctors

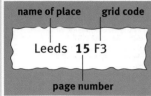

name of place   grid code

Leeds **15** F3

page number

World Flags

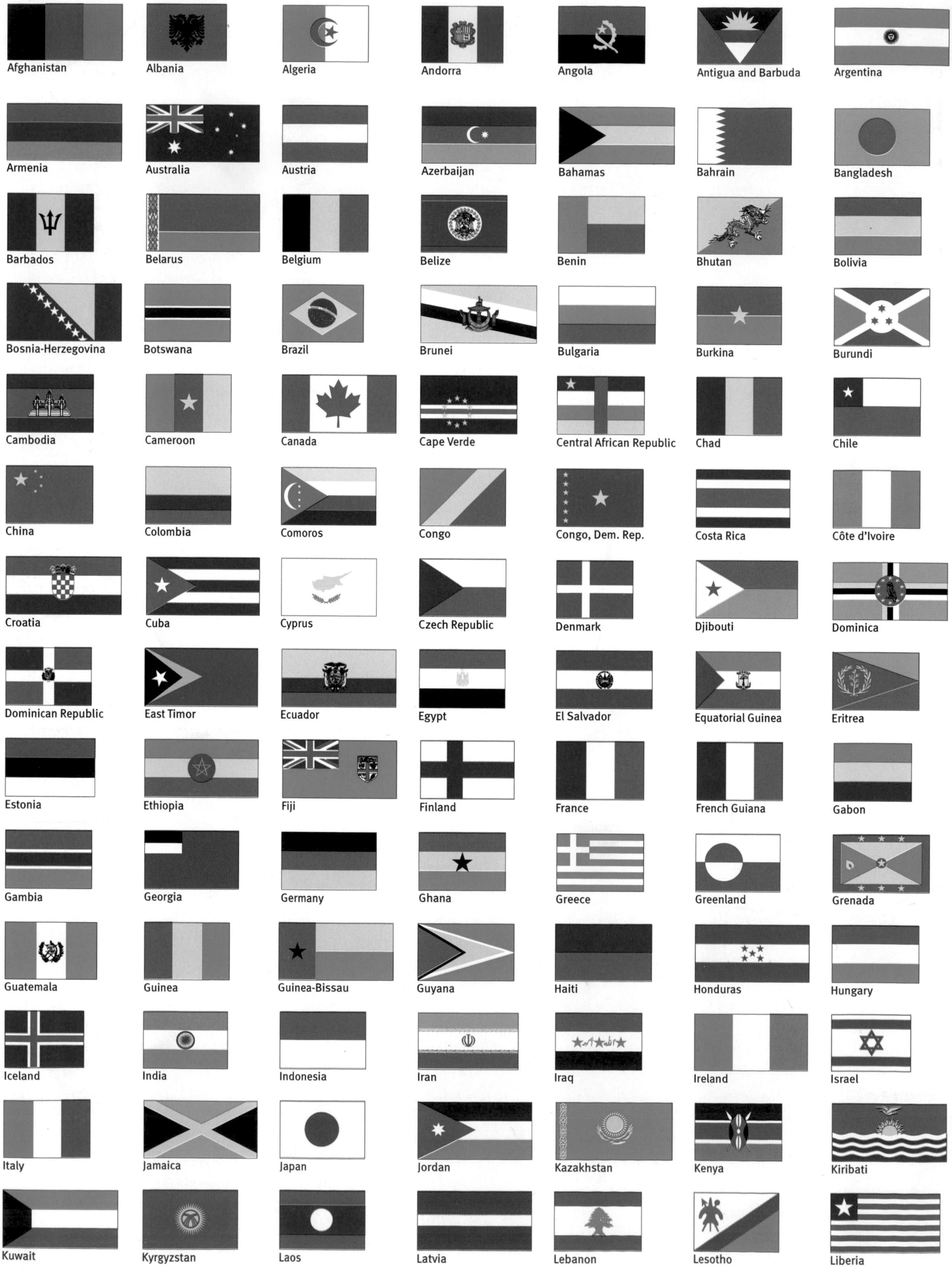

Afghanistan
Albania
Algeria
Andorra
Angola
Antigua and Barbuda
Argentina
Armenia
Australia
Austria
Azerbaijan
Bahamas
Bahrain
Bangladesh
Barbados
Belarus
Belgium
Belize
Benin
Bhutan
Bolivia
Bosnia-Herzegovina
Botswana
Brazil
Brunei
Bulgaria
Burkina
Burundi
Cambodia
Cameroon
Canada
Cape Verde
Central African Republic
Chad
Chile
China
Colombia
Comoros
Congo
Congo, Dem. Rep.
Costa Rica
Côte d'Ivoire
Croatia
Cuba
Cyprus
Czech Republic
Denmark
Djibouti
Dominica
Dominican Republic
East Timor
Ecuador
Egypt
El Salvador
Equatorial Guinea
Eritrea
Estonia
Ethiopia
Fiji
Finland
France
French Guiana
Gabon
Gambia
Georgia
Germany
Ghana
Greece
Greenland
Grenada
Guatemala
Guinea
Guinea-Bissau
Guyana
Haiti
Honduras
Hungary
Iceland
India
Indonesia
Iran
Iraq
Ireland
Israel
Italy
Jamaica
Japan
Jordan
Kazakhstan
Kenya
Kiribati
Kuwait
Kyrgyzstan
Laos
Latvia
Lebanon
Lesotho
Liberia

 Libya

Liechtenstein

Lithuania

Luxembourg

Macedonia, FYRO

Madagascar

 Malawi

Malaysia

Maldives

Mali

Malta

Marshall Islands

Mauritania

 Mauritius

Mexico

Micronesia

Moldova

Monaco

Mongolia

Morocco

 Mozambique

Myanmar

Namibia

Nauru

Nepal

Netherlands

New Zealand

Nicaragua

Niger

Nigeria

Northern Marianas

North Korea

Norway

Oman

Pakistan

Palau

Panama

Papua New Guinea

Paraguay

Peru

Philippines

Poland

Portugal

Qatar

Romania

Russian Federation

Rwanda

St. Kitts and Nevis

St. Lucia

St. Vincent & the Grenadines

Samoa

San Marino

Sao Tomé and Principe

Saudi Arabia

Senegal

Serbia and Montenegro

Seychelles

Sierra Leone

Singapore

Slovakia

Slovenia

Solomon Islands

Somalia

South Africa

South Korea

Spain

Sri Lanka

Sudan

Suriname

Swaziland

Sweden

Switzerland

Syria

Taiwan

Tajikistan

Tanzania

Thailand

Togo

Tonga

Trinidad and Tobago

Tunisia

Turkey

Turkmenistan

Tuvalu

Uganda

Ukraine

United Arab Emirates

United Kingdom

United States of America

Uruguay

Uzbekistan

Vanuatu

Venezuela

Vietnam

Yemen

Zambia

Zimbabwe